UNDERSTANDING YOUR GARDEN

*The science and practice of
successful gardening*

Stefan Buczacki

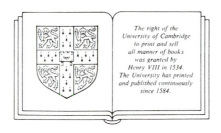

The right of the
University of Cambridge
to print and sell
all manner of books
was granted by
Henry VIII in 1534.
The University has printed
and published continuously
since 1584.

CAMBRIDGE UNIVERSITY PRESS

Cambridge
New York   Port Chester
Melbourne   Sydney

Published by the Press Syndicate of the University of Cambridge
The Pitt Building, Trumpington Street, Cambridge CB2 1RP
40 West 20th Street, New York NY 10011, USA
10 Stamford Road, Oakleigh, Melbourne 3166, Australia

First published 1990

Printed by Kim Hup Lee Printing Co, Singapore
Designed by Chris McLeod

British Library cataloguing in publication data

Buczacki, Stefan
Understanding your garden.
1. Gardening
I. Title
635

Library of Congress cataloguing in publication data

Buczacki, S. T.
Understanding your garden: the science and practice
of successful gardening / Stefan Buczacki.
         p.   cm.
Includes index.
1. Gardening.   2. Garden ecology.   3. Botany.
I. Title.
SB453.B874   1990
635–dc20   90-38131   CIP

ISBN 0 521 33468 3

GO

CONTENTS

CHAPTER ONE

Gardening with science

Why should a gardener be concerned about science? What can science offer to make gardening more successful or enjoyable? The reader will not be surprised if I answer 'a great deal', but I believe that it is simply misunderstanding of what science is and what science does that prompts such questions. For even now, in the late twentieth century, there is something of an aura about science.

We live, so we are told, in the age of science and technology, yet anyone describing their profession as 'scientist' still attracts just the merest hint of a raised eyebrow, just the slightest quizzical expression and just the very tiniest indication of wariness on the part of their companions. Even when science is applied to gardening or horticulture, the reaction tends to be one of some interest in the explanation that science offers for an established gardening happening or technique, yet usually suggests a polite hint of caution and diffidence towards a novel procedure that is actually based on scientific principles and reasoning.

The historical attitude of wariness, fear and even hostility towards people who had explanations for natural things beyond the ken of most of their fellows is well known and was once allied to their real or imagined distrust of deeply held religious beliefs. Today, religious worries about science have largely been supplanted by ones allied either to man's innate aggression towards his fellows (nuclear warfare) or to his general lack of concern for other living inhabitants of the planet (environmental contamination). Admittedly, it is science and scientists that have provided the wherewithal for all of these things, but so it is science that has offered to much of mankind a better, safer, easier and healthier life.

Superficially, a good and attractive garden appears almost self-sufficient although it takes much care, attention and experience to achieve such an effect. Underlying every facet of gardening activity and experience is at least one, generally unappreciated, scientific principle.

Even in the smallest of gardens, careful planning is needed to achieve the best results. In many ways, it is by working within a limited space that the gardener most appreciates some knowledge of the underlying scientific basis of his or her actions.

This book has a scientific basis, yet it is not about warfare and only very slightly touches upon environmental contamination and public health. My purpose is embodied in the etymological roots of the word science – in the Latin *scientia*, knowledge, and *sire* to know. Scientists seek after knowledge – knowledge of the physical and material universe – and, in large measure, it is society's judgement that dictates the use that is to be made of the knowledge. Scientists acquire knowledge in three main ways: by observation, by measurement and, above all, in the sort of science that concerns us here, by experiment.

An experiment is usually performed to answer a question and usually to test the validity of a hypothesis or hunch of what the answer might be. Take, for instance, the problem of obtaining the maximum number of onions of a certain size from a given area of garden. The experimenter might hypothesise that the answer lay mainly in choosing the appropriate variety but secondarily in applying fertiliser at a particular time and thirdly in the spacing at which the seeds were sown or the plants planted. He or she would therefore devise an experiment that tested several values for each factor – several varieties, fertilised in different ways

Scientific experimentation can take many forms. Here, in an entomology laboratory, studies are being made of the manner in which some plants are rendered more or less attractive to insects than others.

These cages, with nets of varying opacity, are used to investigate the effect of different light intensities upon the development of plant disease symptoms.

and spaced to different distances. Of course, the experimenter's own experience would be invaluable in deciding which of the hundreds of possible varieties should be tested and which of the almost limitless permutations of fertiliser and spacing would most probably be relevant. The experiment could offer several answers. Perhaps the choice of fertiliser made no difference at all and perhaps some varieties responded to spacing differences whereas others did not. Every experimental result suggests a further experiment or series of experiments, but in each case the result must be tested, for it is a crucial feature of the scientific method that the findings must be repeatable. A particular finding could otherwise be due to some additional complication – the weather or the sowing date in our onion experiment, for example.

In modern experimental science, a statistical test is almost always applied to the findings and all scientists today are likely to have a working knowledge of statistical probability. This is the discipline that applies a mathematical test to calculate the relative likelihood of an observed result being due to chance. Or, to express it another way, to estimate the degree of confidence that one may have in the findings being valid, on a scale ranging from zero (impossibility) to one (certainty). In the onion experiment, for instance, a statistical test would indicate if there was a one in ten, one in a hundred, one in a thousand or even smaller likelihood of the results having arisen fortuitously. Clearly, other things being equal, the greatest weight, value and importance attaches to a result that has only about a one in a million possibility of having arisen by chance.

Having obtained their findings (and, today, tested their validity statistically), scientists attempt to formulate rules or laws to codify their observations. Their success is often to be measured in how widespread and generally applicable are these laws or explanations. Ask a layman (or even a professional scientist) to name the greatest scientific practitioners and the list will inevitably include da Vinci, Newton and Einstein, all of whose findings are of universal application. Gravity, for instance, is always gravity and obeys the same laws wherever the force is applied. In biological science, Darwin and, in our time, Crick and Watson stand out as scientists whose names are familiar to the public at large. They all produced observations or experimental results or deductions that were of very general application and relevance.

In a garden, I like to think of science's contribution being either direct or indirect. Within the contributions of indirect science, I include plant breeding, entomology and plant pathology. Few gardeners indulge in hybridising wild plant species in order to attempt to improve the yield of vegetables or investigate the life cycles of aphids or evaluate potential new fungicides. But whenever we grow a vegetable, control an aphid infestation or spray our roses to control blackspot, we are indirectly benefiting from the scientific advance that someone else has made. Consider also a plant in its plant pot, a situation encountered by almost

every gardener every time he or she gardens. Unthinkingly, gardeners benefit from the technology of the plastic plant pot, the formulation of the potting compost, the type of fertiliser it contains, the process used to sterilise the loam and, of course, from the hybridising, selection, seed production and packaging that enabled them to raise the plant. I believe there is much interest and satisfaction to be derived from knowing of these things and of understanding why something we do has been made, prepared or suggested in a particular way.

However, there are also ways in which science has improved directly the way that gardening is done. What is the best way to dig, the most efficient way to sow seeds to ensure germination, the most assured manner of storing vegetables or of watering the garden? What is the basis for choosing particular fertilisers for particular tasks or certain weedkillers for certain problems? I believe that all gardeners ultimately will be interested in knowing how science can help them to make up their minds over which option to choose, which course to follow and which way to do a particular gardening task.

It is with these beliefs in mind that I have written this book, but two further points should first be made. Inevitably I have drawn many of my examples and findings from commercial horticulture for it is a maxim of life that scientific research costs money and is only likely to be expended in areas that are themselves contributors to national revenue. However, I have adapted and extrapolated this research to the garden situation wherever possible. And I am often asked about the effectiveness or validity of so-called traditional techniques and beliefs – including those sometimes referred to as old wives' tales in which gardening abounds. How do I square my scientific training and approach with these seemingly unscientific principles? Quite simply – I do not spurn them or reject them just because science has no explanation for their effects, because the right observations, measurements and experiments have not been made. This is no basis for stating that the suggestion is false, merely that it cannot yet be proved.

The clinostat (left) is one of the first items of experimental equipment to be encountered in school biology classes. It is commonly used to demonstrate the effect of gravity on plant growth. With gravity removed by rotation, the test plants will grow horizontally, unlike the static plants seen on the right.

Another common school experiment demonstrates the importance of temperature in bringing about seed germination. The pea seedlings on the right have been incubated at the optimum temperature while those in the centre were kept too hot and those on the left too cool.

CHAPTER TWO

What is a plant?

Plant life takes many forms and occupies many different types of habitat. Even within the restricted area of this garden, it is possible to see great variation in leaf form, flower shape and size, overall plant size and form, and relative preference for wet and dry areas.

In schools and colleges today, even at an advanced level, students are taught biology, the science of life, rather than the two disciplines of botany, the study of plants, and zoology, the study of animals. This approach has its drawbacks but at least it has eliminated the dreadful and quite unreasonable schism that once existed and which resulted in botany being considered by many people as the tedious, boring and effete subject of the two. Nevertheless, whilst animals may move and have fairly obvious personalities, the plant is in many ways a much more exciting organism. It displays quite astonishing diversity in overall form and has adapted the same basic methods of living and feeding to almost all of the environmental extremes that our planet has to offer. As gardeners we are privileged to have plants as our raw material. In modern biological science, however, it is recognised that some living things cannot be classified as either plants or animals for, like fungi, bacteria and viruses, they are so different as to warrant having their own kingdoms. I shall be referring to these later in the book when I discuss their role in the garden. For the present, however, I shall concentrate on the two major divisions.

Plants and animals, of course, are living organisms. They differ from non-living things in their abilities to grow, feed and reproduce themselves, and it is to satisfy the plant's basic functions of feeding and growing, and to some extent of reproducing, that gardeners garden. Plants differ from animals most fundamentally in the way that they feed; animals are dependent upon some external source of food in the shape of other organisms, whilst plants are able to utilise non-living materials and the energy derived from the sun. Animals are also usually capable of locomotion – movement from one place to another – whereas plants are not. Animals grow to a definite size, with a defined number of individual organs (one heart, two lungs and ten finger-nails in human beings for instance), whilst plants have more or less continuing growth to a more or less indefinite ultimate size.

Most of the plants that occur in gardens produce seeds and most of them bear or are potentially capable of bearing flowers; they belong to the large group called flowering plants or angiosperms. However, gardens often contain a good representation of seed-producing but cone-bearing plants (usually called conifers or gymnosperms) too. Angiosperms are divided into two large subdivisions: the monocotyledons which have one cotyledon or seed leaf and the dicotyledons which have two. Monocotyledons usually have narrow, strap-like leaves – grasses and most bulb- and corm-forming plants belong to this group. Collectively, angiosperms and gymnosperms are known as higher (meaning more advanced) plants, but some gardeners grow a few types from among the so-called lower plants too. Ferns are the best known amongst these, although the more unfortunate gardener may also encounter their relatives, horsetails, growing in gardens as almost ineradicable weeds. These lower groups do not bear seeds but reproduce instead by means of much simpler structures called spores. Even simpler spore-producing plants are found in gardens too – mosses and liverworts on the lawn and paths and algae in the garden pool for instance. We now believe that the flowering plants evolved from spore-bearing types and first appeared on the earth about 100 million years ago.

A flowering plant has three major criteria to satisfy if it is to function properly. First, it must have some means of supporting itself physically so that the tissues containing light-sensitive chemicals are exposed to the sun in the most efficient way possible. Second, it must have a means of supporting its reproductive structures and exposing them in a position where fertilisation and seed maturation and dispersal can be performed. And third, it must possess a system that permits the easy movement of water and nutrients into and around its various parts.

The microscopic structural unit that provides a plant with the wherewithal to achieve these three ends is called a cell. A typical plant cell may be thought of as a tiny, irregularly shaped bag of watery fluid or protoplasm containing a 'nerve

Among the many non-flowering plants common in gardens are liverworts, of which this species of Marchantia illustrates the flattened, leaf-like thallus from which a reproductive body arises. Unlike their close relatives, the mosses, liverworts are seldom troublesome as weeds.

everyday and so conspicuous that it is even remarked upon by astronauts as they return from space. Many plant cells therefore are green and they contain chloroplasts even when photosynthesis is subordinate to a structural or other function. But beneath the outer surface or epidermis of leaves lie tissues for which photosynthesis is the primary role. The concentration of chloroplasts is greatest (as many as one hundred per cell) in the tissues closest to the epidermis and also on the side of the leaf facing the sunlight, for not only does chlorophyll trap solar energy but light is also required for the chemical itself to be formed. Hence, plants kept in the dark turn yellow because no new chlorophyll forms and they die in consequence because no food can be manufactured.

A glance into your garden will tell you that chlorophyll is not the only pigment that plants contain – there is very little green coloration to beetroot leaves or roots, to a dahlia flower, to the fruits of *Rosa rugosa* or to the leaves of maples in autumn. Sometimes, as in flowers, chlorophyll is absent from the tissues and the colour derives largely from pigments called anthoxanthins, anthocyanins and carotenoids. Sometimes, as in beetroot leaves, chlorophyll is present but masked by red anthocyanin of uncertain function.

Physical strength to support leaves and flowers can be imparted to a plant in two main ways, examples of both of which are familiar to gardeners; and both depend on features of the cells. One method relies on thin-walled cells that are inflated because of the pressure provided by the watery liquid cell contents inside. (Imagine a balloon filled with water to gain an idea of the strength that can be supplied to an otherwise fragile object in this way.) Understandably, this system is dependent on a continuing external supply of water to the plant and as soon as this is denied or the cell punctured, the pressure and the structural strength is lost – hence the soft parts of a plant starved of water will wilt. Because of the vulnerability of the cells to puncture or dehydration, the water-inflated cells are usually given added protection by a toughened

centre' or nucleus and other minute organs or organelles confined by a boundary or wall. Cells are rarely larger than 50 micrometres or smaller than 10 micrometres in any dimension but do exhibit great variety both in size and shape. Some unsophisticated forms of algae comprise one cell only, but all other green plants are made up of aggregate cells called tissues. Within each tissue, the cells tend to be of similar form and have a specialised common function that complements the specialised functions of other tissues and enables the plant as a whole to grow.

The most important and at the same time the most interesting specialised plant cell function is the ability to photosynthesise, to use the energy in sunlight to convert carbon dioxide and water vapour from the air into carbohydrates. Cells that are able to photosynthesise contain structures called chloroplasts – more or less discus-shaped objects usually having a diameter of about 3–4 micrometres. The chloroplasts contain chlorophyll, a green pigment that traps the energy in sunlight to enable the chemical reactions of photosynthesis to take place, and it is this one chemical that gives our planet its predominant colour, seen by all of us

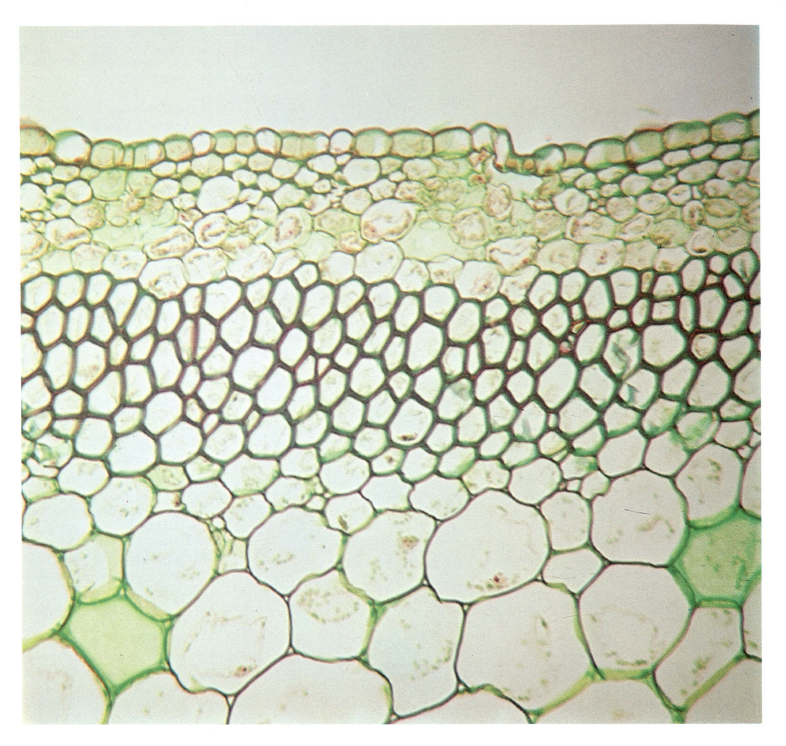

epidermis or skin which limits water loss. The value of the toughened skin in protecting turgid or swollen cells is seen very evidently when the surface of a peach or tomato is punctured. All plants gain at least some structural strength by this method which is called a turgor system (turgor is a word derived from the Latin for swelling), but apart from a few succulents such as species of *Aloë*, the desert stone plants (*Lithops*) or saltmarsh species like *Salicornia*, few rely on it for the majority of their structural strength once they have developed beyond the seedling stage.

Some device more robust than mere water pressure is needed to support the great bulk that many flowering plants and conifers attain and this is provided by cells with their walls specially thickened. This thickening is sometimes provided by extra deposits of the cellulose that all mature plant cell walls contain. (Cellulose is a chemical polymer, comprising many glucose molecules attached together in chain fashion.) Such cells occur typically in young stems or other parts still actively growing and elongating, for although physically strong, they remain fairly flexible. Alternatively, cell walls can be strengthened by the formation of complex chemical polymers called lignins. Once their walls become lignified, cells usually die, but they still impart a plant's main structural strength – witness the fact that even dead trees do not collapse but remain standing, starkly upright. The lignified cells of mature trees and shrubs are largely concentrated in the huge central mass of dead water-conducting tissues that we call wood. Even in young plants the water-conducting cells and some of those closely associated with them are lignified although their structural significance at that early stage of a plant's life is small compared with that of the cells having extra cellulose.

All flowering plants can be divided into several more or less distinct parts – the roots, stem, leaves and flowers, the latter ultimately giving rise to seeds and sometimes fruit too. All of these structures or organs are composed of the same basic range of cell and tissue types although the cells and tissues are arranged in different ways and in different proportions to satisfy each organ's specific role.

Roots have two main functions: they provide physical anchorage for the remainder of the plant and they provide the means for plants to take up water and mineral nutrients from the soil. In addition, and in common with stems and other plant organs, they can be modified to provide certain other functions, most notably food storage – which is why we take so much trouble to cultivate parsnips, beetroot and turnips for their roots alone. (It should be appreciated that not all underground parts of a plant *are* roots, for stems and buds can be modified for underground growth too, and at least one familiar plant, the peanut or groundnut, actually buries its flowering head below ground after fertilisation in order for the fruit to mature in the soil.)

Although gardeners talk routinely of 'the roots', this is a simplification and can mislead; pull up a carrot, a tuft of grass and a mature oak tree and you will see a little of the variation that roots can display. The long, tapering main root of the carrot is called a tap root and this continues to grow and elongate throughout the plant's life. The carrot is a biennial, but long-lived perennials, even herbaceous perennials like horse-radish, can easily develop a tap root to depths of a metre or more on suitable soils. Tap roots of this type are usually unbranched although shorter, stout and rather sparsely branched tap roots also occur. In most monocotyledons, like grasses, and in many dicotyledons too, the branch roots comprise virtually the entire root system, ramifying extensively but relatively shallowly near the soil surface. The oak, however, exemplifies those plants that have a particularly efficient rooting system, beginning as a seedling with a single tap root, progressing as a sapling through the development of short lateral roots, forming thick branch roots after about ten years and eventually having an extensive system of very stout branch roots from which sinkers (in effect, further tap roots) descend to great depth.

The physical strength required by roots is rather

The key to much understanding about plant structure and function is the microscope. This microscopic section, or slice through a plant stem, reveals that the basic structural units, the cells, vary greatly. Some have thickened walls for protection against dessication or to impart structural strength, while others are larger and thin-walled, permitting ready diffusion of substances in solution from one to the other.

special for they must be able to withstand straining and pulling forces (as when the wind threatens to wrench a plant from the ground), but because of the constraints of the surrounding soil, they do not have much requirement for bending or flexing. When a root is sliced through, therefore, and its internal structure examined, the main strengthening component in the shape of the conducting tissue will be found towards the centre with a large mass of flexible tissue around it. This imparts a rope-like quality, rather different from that of a stem.

The nutrient and water uptake functions of roots are not performed by the main root structure itself but by small, almost microscopic extensions of some of the outermost cells. These tiny extensions are called root hairs and they occur on almost all roots, concentrated in a small area a few millimetres back from the tip. The most familiar among those plants that lack root hairs are the aquatic species. They take up very little water (having no need to do so since they live in a stable environment where evaporation and water loss do not occur) and absorb nutrients through their leaves. When normal, non-aquatic plants grow in water-logged soil, one reason why they so often develop unsatisfactorily is through their functions being impaired by inadequate nutrient uptake consequent on poor root hair formation. Root hairs *en masse* can be seen like a tuft of cottonwool when a young seedling is uprooted or when seeds are allowed to germinate in moist air. They are not only small, but also fragile and usually short-lived, shrivelling away as the root extends and as they become spatially further from the growing tip. It is primarily the fragility of root hairs and the damage caused to them when plants are uprooted that makes it difficult to transplant successfully in the height of summer – plants moved at such times continue to lose water through their leaves while being unable to replenish it from below.

The root system of a mature plant is very extensive (it is a common gardening maxim to assume that there is as much of a tree below ground as

there is above it), and those patient scientists who have measured total root systems have produced such astonishing statistics as the fact that in a mature vegetable plot there may be 20 kilometres of root beneath every square metre, or, even more incredibly, that a single rye plant can produce over 600 kilometres of root. But whilst a branch (or, as it is sometimes called, fibrous) type of root system has the merit of being able to reach nutrient and food resources throughout a large volume of soil, it does not have the anchoring efficiency of a tap root. This is demonstrated by the relative ease with which many large trees like beeches are blown over in gales whilst even a modestly sized horse-radish plant is uprooted with the greatest difficulty by the strongest man.

The modification of roots for food storage is a feature of many different types of biennial and perennial plants, providing them with the means of survival during a period when their above-ground parts die down. The carrot, mentioned earlier, has a modified and swollen tap root for this purpose, but swollen branch roots are commoner and are usually called tubers. Dahlias, celandines and Jerusalem artichokes are familiar examples of tuber-producing plants and the actual swelling is brought about principally and usually by an increase in the numbers of thin-walled cells in the root cortex (the area external to the conducting tissues) although other root tissues are modified for storage in some plants. The chemical form in which nutrients are stored varies from species to species, but all storage organs are inherently soft and fleshy and prone to rotting after fungal or bacterial infection.

Roots are developed above ground by a few plants, especially in moist tropical environments where many species (orchids, for instance) grow on the trunks and branches of trees (an epiphytic habit), their roots dangling into, and taking water from, the moist air. Other plants, even in temperate climates, benefit by having some root development from the stem just above ground level but this differs from the aerial roots of most tropical species in being almost entirely for support.

The waterlily is the most familiar example of a plant that is adapted to an aquatic way of life. Its stem is abbreviated to form a crown from which roots pass downwards to anchor in the mud of the pond, while the leaves and flowers rise to float on the water surface.

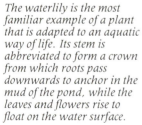

Among the most extreme root modifications are those found in such climbing plants as the ivy. The aerial roots take on a grasping function, adhering by tiny suckers to flat surfaces.

It is a common feature of large monocotyledons like palms, which have massive, unbranched and inherently unstable stems and no stout tap root in the soil. In a few so-called self-clinging plants like ivy, modified aerial roots actually have an adhesive function, anchoring the stem to smooth, vertical surfaces.

Unlike the root, with its dual roles of anchorage and water/nutrient uptake, the stem has primarily a structural function, holding the leaves and flowers above ground, although it also acts as a channel or pipe through which both food and water pass on their way up and down between roots and leaves. (Water and mineral nutrients are transported up through the plant in xylem tissue, whilst the

nutrients made during photosynthesis move through the phloem tissue.) In addition, most stems, at least when young, are green and as capable as leaves of photosynthesising although their overall contribution in this regard is much more limited. Stems also possess pores or stomata for the control of water loss, but again these are usually much less significant numerically than those of leaves.

As in roots, the structural strength of stems derives largely from the presence of lignified tissues, but the structural requirements of the stem are rather different for it must be pliable to resist the force of the wind which threatens to break it. As any engineer knows, a tube resists such force

A woody structure imparts the one enormous advantage to those plants that possess it, that they are able to remain above ground all year round. In spring, therefore, they have a head start over herbaceous species growing afresh from ground level.

much better than a rod of similar cross-section and so, unlike the central lignified core of conducting tissues found in roots, stems (at least when young) have their strengthening components arranged in a cylinder. Usually this is merely a ring (or, in mono-cotyledons, several irregular rings) of isolated conducting and other lignified tissue elements, but eventually (and generally when radial growth has ceased) can be a complete internal tube. Only in older trees does the stem become something of a solid rod, although by that stage the combination of root anchorage and overall robustness means that, although only slightly flexible, the plant can tolerate all wind strengths short of a gale. Moreover, whilst roots are almost always more or less circular in cross-section, many plants have stems that are characteristically angular, the sharp corners and associated structural strength being brought about, as in the dead-nettle family, or Labiatae, by lignified cells in the angles.

But how does the change occur from a young tree seedling, with a few discrete bundles comprising lignified water-conducting cells (xylem) and non-lignified food-conducting cells (phloem) arranged in a ring towards the centre of its stem, into a mature plant in which a cross-section of the stem reveals lignified tissue almost to the exclusion of everything else? It occurs by means of a process called secondary thickening in which new conducting tissues are first formed between the discrete bundles to produce a complete cylinder in the centre of the stem. Subsequently, more and more lignified water-conducting tissue is produced each year to the outside of this cylinder. Gradually, the dead water-conducting cells towards the centre become filled with the gum-like substance resin or other deposits. In the mature tree or shrub, therefore, there is a central mass of woody tissue providing purely physical support. Around it are lignified cells that have the dual purpose of support and water transport and, outside them, thin-walled, non-lignified, nutrient-conducting cells restricted to a thin layer just below the outer protective region called bark. The annual additions to the wood give rise to the familiar pattern of rings seen when a temperate climate tree is felled. (Tropical trees, growing in a more or less uniform year-round climate, display very little ring development.) Study of the width and other features of these rings can reveal much about the season to season changes in growing conditions and about climatic fluctuations in past ages.

Timber is the functional product derived from the unique properties of the complex chemical lignin. Lignin imparts structural strength to large plants and forms wood. Very few types of plant have been able to develop to tree size without the ability to form wood through the process known as secondary thickening.

However, not all mature plants that require the additional support conferred by lignin are trees, or indeed contain any wood in the popularly accepted sense of the term. Yet there is no denying the strength and toughness of the stems of *Phlox*, Michaelmas daisies and other herbaceous (or non-woody) perennials that gardeners cut down in vast quantities every autumn. Here the structural strength derives in part from the lignified cells of the water-conducting tissue but often from elongated lignified cells that occur outside the central ring of conducting tissue bundles. Monocotyledonous plants like bamboos lack true secondary thickening and real wood but nonetheless attain a size greater than that of many dicotyledonous trees. They usually do so by an increase in the total number of conducting tissue bundles. However, despite the general efficiency in wild plants of support methods which do not involve secondary thickening, a gardener still finds it necessary to stake such herbaceous perennials as dahlias because many generations of artificial breeding and selection for types having horticulturally desirable large blooms has resulted in plants that are physically unstable and incapable of being held upright by their own stems.

The way that stems branch is curious for, almost invariably, seed plants produce branches from buds in the axils of their leaves (that is, from the angle between a leaf and the stem), and from nowhere else. (The position on a stem where leaves arise is called a node.) Not all leaf axils necessarily produce a branch and some produce more than one, but when a branch does arise, this is where it originates. The frequency of leaf axils bearing branches, together with the pattern or arrangement of the leaves on the stem (in opposite pairs, in threes or offset in some way) therefore dictates the overall form of the whole plant. So, unwittingly, naturalists and gardeners make considerable use of branching patterns when identifying plants from their overall form. There are two other features that aid them in this – features that are best seen with deciduous trees in winter – the branching angle and the degree of branching. The branching angle may range from very small, with the branches almost upright, giving a columnar or fastigiate appearance to the plant, to very large, when the branches actually hang downwards and the whole structure is called pendulous or weeping. Of course, most plants fall between these two but, interestingly, the same species can sometimes occur in both extremes. The hawthorn, *Crataegus monogyna*, for instance, occurs as the weeping form 'Pendula' and also the fastigiate 'Stricta'. The degree of branching also varies widely, especially among perennials, from the single wholly unbranched stem of most monocotyledons, seen most dramatically in the tree-like palms, through the rather limited branching and thick twigs displayed by most tropical dicotyledonous trees, to the highly repeated branching of temperate climate species whose branches end in multitudes of thin twigs.

The length and form that stems display varies enormously and gardeners will be familiar with examples from both extremes. The duckweed (*Lemna sp.*) that covers the garden pool has to all intents and purposes no stem at all, while the Douglas fir (*Pseudostuga menziesii*), a fairly common tree in larger gardens, is in its native North American habitat among the world's tallest plants – one reputedly of 128 metres was felled in 1895, and, at over 50 metres, it is certainly the tallest species of plant growing in Britain.

Some of the special purposes for which stems can be modified are similar to those of roots. Most climbing plants manage to climb by the twisting action of a slender, pliable stem (as in runner beans) or by modified grasping stem structures called tendrils (as in grapevines). (However, not all tendrils are in fact stem structures (see p. 22). Another common modification is the reduction of entire shoots to the form of spines, partly for protection and partly to minimise water loss. Hawthorn is an example of a plant with stem spines, and it forms an interesting contrast with the superficially similar gorse whose spines are modified leaves and also, of course, with cacti, which have

their leaves reduced to spines, but borne on a swollen stem. No one gardens for long before becoming familiar with those stem modifications that serve to facilitate plant spread. The runners of strawberries, for instance, are slender stems growing along the surface of the soil, producing buds at intervals from which more stems grow upwards and from which roots grow downwards. The offsets of plants like *Sempervivum* are abbreviated forms of runner. The arching stems called stolons that are formed by blackberries root where they touch ground, while the underground runners called suckers can arise, as in roses and plums, from a bud on a root or, as in mint, from an underground stem or rhizome. The ease with which damaged root tissues tend to form buds from which suckers arise is soon discovered by the gardener who is ill-advised enough to attempt root pruning of plums. And the efficiency of the rhizome as an organ of spread is nowhere better illustrated than in couch grass (*Agropyron repens*), one of the most troublesome garden weeds but one that actually produces few flowers and few seeds, relying almost entirely on its ability to regenerate rapidly from very small rhizome fragments which grow swiftly and deeply to great length.

Extensive modification for food storage is fairly unusual in conventional above-ground stems. This is simply because many of the plants requiring this form of food reserve do so to enable growth to re-start swiftly after a period of adverse conditions (in the spring after a temperate climate winter for instance) and the food store itself would be vulnerable to damage during the dormant period. A few types of plant, like the desert cacti, have stem tissues modified for water storage, but in general food storage in stems is restricted to those types of stem that have adapted to exist *below* rather than above ground. Superficially, many of these underground structures (bulbs, corms and rhizomes, for instance) appear similar to modified roots although the fact that they are at least in part composed of stem tissue is revealed externally by a terminal bud and by the presence of (or the ability directly to

Spines may be modified leaves or, as in this rose, modified parts of the stem. Their role is not simply to cause discomfort to gardeners but to serve as grasping structures and so support the plant when, in natural conditions, it scrambles over other vegetation.

The leaf is often a beautiful structure aesthetically, but in its photosynthetic and water regulatory roles it is also a beautiful example of the harnessing of form to function. The typical leaf represents a compromise between providing the maximum surface area for sunlight capture and minimising water loss to the point where it adequately helps to draw up water from below but does not deplete the plant of moisture more rapidly than the roots can balance with uptake from the soil. The compromise is dictated largely by the climate and microclimate of the plant's native habitat, and that it sometimes fails is demonstrated by large-leaved plants wilting in a dry soil in the height of summer and evergreens wilting in midwinter when the soil water is frozen and inaccessible.

Although most leaves are subdivided into a stalk (even if this is often rudimentary) and a more or less flattened blade, the variation in overall leaf form among plants is enormous and is utilised extensively in gardens for its visual effect. The simplest leaf form is called lanceolate and its overall appearance is that of a spear-head, but from this basic type variations occur that encompass all manner of intricate and complex indentations, lobes and teeth. Many leaves are subdivided into separate leaflets in a range of different ways but almost all of these shapes are basic adaptations to the two purposes of light capture and water regulation in different habitats. I believe that the appearance of the leaves tells a gardener more than any other morphological feature about the conditions that any particular plant is likely to require in the garden. For instance, the highly indented leaf of a woodland fern provides a large surface area for light capture in a poorly lit environment, whilst its smooth surface and overall thinness, which would encourage water loss, are indicative of a moist atmosphere in which such loss would present no hardship. But place this fern in direct sunlight and the fragile leaf tissues will almost certainly scorch and the plant suffer as its photosynthesising tissues are inevitably dried out. Conversely, the reason that so many alpine plants are difficult to grow in

form) leaves. To confuse matters further, the same word, tuber, is used for both root and stem structures, but root tubers like Jerusalem artichokes are leafless whereas stem tubers like potatoes possess tiny scale leaves.

The leaf is arguably the most significant and most interesting structure in all of botany. As I have mentioned already, it is the main organ containing the chlorophyll that provides the green plant with the means to manufacture basic carbohydrates from gaseous raw materials. The leaf is less generally appreciated for its role in the water relations of the plant, although the pores or stomata through which water loss or transpiration takes place lie on the leaf surfaces. It is sad that so few gardeners ever have the opportunity to see the structure of stomata through a microscope for their simplicity and efficiency of functioning are remarkable. Punching the impervious cuticle covering the leaf surface, each stoma comprises an opening surrounded by two guard cells that change their shape and volume in response to the moisture content of the atmosphere, thereby closing or opening the aperture (p. 37).

gardens is because they have densely hairy leaves, adapted to minimise water loss in the drying mountain-top conditions of high sunlight and strong winds. Those same hairy leaves in a lowland garden will trap moisture and encourage the setting in of fungal decay.

Because of the overwhelming importance of leaves to a plant for photosynthesis and water regulation, it is relatively rare that they display significant modifications for other roles. Even in the vegetables that are grown specifically for the nutrient content of their edible leaves there is rarely any major modification for food storage such as occurs with most other edible plant parts like stems and roots. Perhaps the most frequent leaf modification that we encounter in gardens is in such plants as *Clematis* where it is the leaf stalks and not the stems that are altered to form tendrils to aid climbing.

Flowers must be the plant organs to which gardeners pay most attention, for without their flowers, many ornamental and all fruiting plants have failed to fulfil their allotted role. But not one gardener in a thousand appreciates the flower for its encapsulation of plant evolution. Yet the flower as we see it today is universally recognised by botanists as representing a telescoped leafy shoot, some of the leaves having become modified to form petals and others stamens. They have also lost their photosynthetic function and the whole structure now serves the sole purpose of ensuring that satisfactory reproduction can occur. But just how plastic a structure the flower represents is apparent in every garden. For example, the double forms of roses or other plants that occur naturally only as singles demonstrate how easily stamens can become petals. It is also apparent that the variation in flower form is prodigious and this in itself is one of the great attractions that gardening holds for so many people. I can do no more than outline the significance of some of the main trends in this variation therefore.

The purpose of flowers is to produce seeds. Any individual flower may contain male reproductive

The true nature of the flower is best seen in members of the magnolia family, such as the tulip tree (Liriodendron), seen opposite. With relatively little imagination, the flower can be envisaged as a telescoped shoot, with leaves and shoot tip modified to form the petals, stamens and stigmas.

The 'flower' which a gardener sees can vary greatly in its complexity. Daffodils, for example, are each single blooms bearing scaly sepals, brightly coloured petals (the inner ones fused to form the trumpet), stamens and stigmas.

The apparently single flower of the daisy family, exemplified by the dahlia, is in fact a mass of individual flowers of differing shape and function. The 'petals' are the outer ray florets bearing male reproductive parts while the 'stamens' in the centre are in reality the tubular female organs.

structures (pollen-producing stamens), female reproductive structures (pollen-receptive stigma and ovaries) or both. And where flower sexes are separate, an individual plant may bear either male flowers or female flowers or both. The sizes and arrangements of flowers are also immensely varied. The daffodil is fairly evidently one solitary entity but most flowers occur in groups, like the cluster on each rose stem. These groups of flowers are called inflorescences. The way that flowers are grouped together into an inflorescence gives much of the diagnostic character and charm to particular plant families – the more or less flat-topped inflorescence called an umbel to the carrot family, the Umbelliferae, for example. The individual flowers in an inflorescence are often barely discernible separately (think of the bloom of a *Buddleia*, for instance), and in one of the largest of all plant families, the daisy family, or Compositae, each apparent 'flower' actually comprises a mass of obscure tiny individual blossoms with the strap-shaped, petal-like males around the periphery and the tubular females in the centre.

Plant fruits vary from such forms as the berry of the Cotoneaster (top), with its seeds covered by a fleshy coat attractive to birds, to the much smaller one of the Geranium which disperses its seeds explosively.

The product of a fertilised flower is a fruit, bearing seeds. Whilst apples, raspberries, plums and black currants are all fairly obviously recognisable as fruits, a little thought will reveal that the numbers of seeds within an individual fruit can vary widely from one in plums and other stone fruits to the multitudes produced by poppies and by many weeds. There are many different types of fruit, named from their differing methods of formation, structure and means of liberating the seeds. Among those frequently encountered in gardens (and not always given their correct names) are legumes (the pods of the pea and bean family), berries (gooseberries, cucumbers, tomatoes – most of the fruits popularly called berries are botanically something else), capsules (the commonest type of all, including poppies, carnations and primulas among many others), drupes (plums, apples, raspberries – the latter actually comprising a group of small drupes) and nuts (hazelnut, acorn). The strawberry exemplifies a particular curiosity in that the objects grown and eaten as strawberries are not actually berries, nor indeed fruits at all, but swollen receptacles (the apex of the flower stalk), and the strawberry pips present over the surface are actually single-seeded fruits called achenes.

I have referred several times to the self-evident fact that plants grow – they increase in size more or less continuously throughout their allotted life span. And I have referred also to the fact that the basic structural units of a plant, as of other living things, are cells of several different types with clearly defined functions. Plainly, therefore, as a plant grows its cells must increase either in size, number or both. In fact, they generally increase relatively little in size, and increase in number only in certain specialised zones called meristems (a word that often causes amusement but which is derived from the Greek for divided). The cells of a meristem divide to produce new cells that, although at first fairly uniform, gradually take on the form, appearance and function of one of the specialisations to which I have referred. This change to a specialised structure and role is called differentiation and later on (p. 101) I shall discuss some of the ways that it is influenced. Meristems occur (and may arise new) in any part of the plant where growth is to take place. The most characteristic meristem of flowering plants is the apical meristem which is present at the tip of each shoot and root and is responsible for its elongation. If the top of a shoot is cut off, therefore, the apical meristem is removed and elongation ceases – the basic theory behind the Dutch hoe as a method of weed control. But any gardener of experience will immediately cite two instances that seem to counter this argument. Some weeds, severed with a hoe, *do* produce new shoots, and lawns continue to grow despite being regularly mown. Those weeds that have had their apical meristem removed yet still manage to grow do so either because an undamaged side shoot takes over the role of the main shoot (see p. 102) or because a new meristem forms in the tissue in response to the action of wounding. (The phenomenon of taking cuttings also exemplifies this ability of plants to produce a meristem and new cells, in this instance root cells, as a result of a wound being formed or tissue damaged.) Grasses present a special case for, whilst they do have apical meristems, increase in height also arises from below – from meristems further down the stem close to soil level, out of reach either of lawnmower blades or grazing animals.

Earlier, I mentioned the way that the increase in girth of tree trunks occurs by the radial approach of new cells, and it is a meristem called cambium occurring circumferentially around the stem immediately under the bark that is responsible for this. The cambium forms new water-conducting xylem cells internally and nutrient-conducting phloem cells externally. And the process of secondary thickening (p. 18), as a result of which wood forms, depends of course on meristematic cells arising in the cortex between the initially discrete bundles of conducting tissue.

Now, with some concept of what is present inside a mature plant, we can begin to consider how the plant reached this state from its starting point of a seed, what influences its growth and development and how the practices of gardening can best be directed towards satisfying its needs.

Plants grow by an increase in the size and more importantly an increase in the numbers of individual cells, although it is obvious that this is far from a uniform process, either from one organ to another or within a single organ over a period of time. The size increase is also rather difficult to measure meaningfully. If the growth of a single leaf is measured by reference to the most obvious feature, its area, an S-shaped curve is generally seen, as in the diagram below, revealing that once the leaf has reached a certain size, area increase stops altogether. Nonetheless, the growth of shoots, the formation of buds and the appearance of other new leaves on the remainder of the plant continues. The total area of all of the leaves on a plant at any particular time is an alternative indicator therefore of the plant's performance, but is exceedingly difficult to measure in practice. Area, moreover, is not a very accurate measure of the performance of a leaf or any other plant organ because each varies so much in shape.

The relationship between leaf area and time is reflected in a characteristically 'S'-shaped curve. This indicates a slow start, a rapid increase in area and then a slowing down to a maximum beyond which there is no further increase.

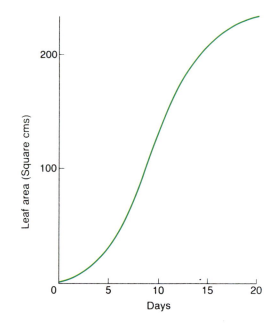

Can we therefore measure plant growth by plotting the increase in the volume of the tissues? Yes, but this is usually very hard to do accurately and can be affected by water content at any particular time. This is also true of measures of fresh weight – anyone weighing the produce from their kitchen garden will know that its weight begins to diminish as soon as it is brought indoors. Scientists generally use the dry weight of plant tissues therefore (drying them for predetermined times at specified temperatures) when they are assessing influences on growth. This is a factor fairly free from the inaccuracies such as I have described although, of course, each plant can only be measured once. Simple measures of height increase are sometimes necessary (foresters, for instance, tend to use this measure for whole trees which are not amenable to being dried) but these can be misleading. As shown below, increase in height of the overall tree or of its main shoot can be very different from the increase in girth (a much better indicator of the volume of timber present) and can give no indication of root elongation. Moreover, increase in height can actually mislead, for a tree in poor conditions (a sunlight-requiring tree growing in deep shade, for example) and actually functioning very badly as a representative of its species may display considerable elongation of its main shoot as it reaches upwards towards the source of illumination.

Why is it that plants grow in such varied ways and how do they manage to respond in area, volume, weight and height to external influences? Why do shoots normally grow upwards and roots downwards? What causes flowers to form in the spring and summer on some species but in the autumn and winter on others? And what dictates that for each type of plant there is an allotted life span? In order to illustrate the main influences and controls on a plant throughout its life, I shall describe very simply some of the factors that govern the development from a seedling to an adult plant and highlight those features that we as gardeners can affect most directly. I cannot do so, however, without introducing a group of chemicals called hormones.

It has long been known that the growth and functioning of higher animals is controlled by chemicals secreted by the endocrine glands – the pituitary, thyroid and so forth. And it was Charles Darwin in 1880 who first realised that plants must contain some comparable 'movable influence'. His observations on the effect of light on grass shoots led him to conclude that whilst the light was actually perceived at the tip, the response to the light was not restricted to the tip but was somehow transmitted to the lower parts of the shoot too. By the late 1920s, further research had established that the 'movable influence' must be chemical and, because of the way that grass shoots had been seen to grow, it was realised that the effect of this chemical must be on cell growth rather than on cell division. Other chemicals of similar effect were discovered subsequently and the general name auxin was given to them. When it became possible to discover the chemical structure of auxins, the original substance was seen to be an organic acid called indoleacetic acid, a name that will probably be familiar to many gardeners in its abbreviated form IAA. For IAA can be synthesised artificially and its ability, not only to bring about stem extension but also root growth, has been used in rooting powders. Both effects arise only at low concentrations of the chemical; as the diagram on the left indicates, at higher concentrations it usually inhi-

The relationship between height, girth and age of an oak tree, showing that while height increase slows down with age, girth continues to rise – hence, ancient oaks tend to be proportionately very 'fat' ones.

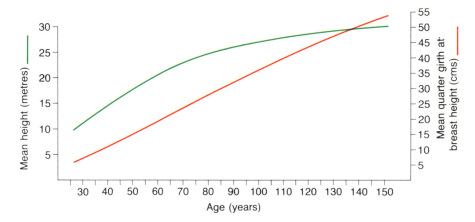

The response of stem and root growth to artificially increased concentrations of the auxin IAA. This highlights the fact that care is needed not to overdose plants with rooting hormone, but also shows that at high concentrations, auxins make valuable weedkillers.

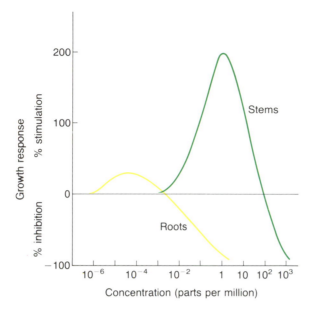

Plants respond in obvious ways to the direction and strength of a light source, but the process of curvature and the lack of chlorophyll are under the control of hormones. Darwin was the first to suggest that the light-induced curvature might have some 'movable' controlling mechanism.

bits growth. Today, most rooting powders actually contain not IAA but derivatives of a chemically related substance, NAA or naphthylacetic acid, which is less readily metabolised or broken down by the plant and so is effective for a longer time. The fact that the growth of different types of plant is inhibited and enhanced at different concentrations of auxin has led to the development of selective hormone weedkillers from some IAA-like substances (see p. 150).

Although many different naturally occurring plant hormones have now been discovered, the most important remain the auxins, the gibberellins (which have a particularly marked effect in stimulating stem elongation in plants that occur as dwarf strains), and the more recently identified kinins. The last-mentioned are of special importance in influencing the rate of cell division; they were first discovered in coconut milk and are characteristically found in embryonic and very young, rapidly growing tissues. But whilst knowledge of the existence of plant hormones has helped us to understand why certain growth effects take place, it must be admitted that the precise mechanisms by which they exercise these influences are elusive.

One of the most obvious types of plant growth response takes us back to Charles Darwin and his observation of the curvature of grass shoots towards the light. All gardeners will have seen many similar effects without perhaps giving them more than a passing thought. Any such type of directed plant growth is called a tropism – green tissues almost invariably grow towards the light and leaves grow such that their greatest surface area is exposed to the maximum illumination. Apparently this occurs because the growth-stimulating auxin is destroyed in cells exposed to the light, growth occurs on the opposite side of the organ only, and the whole bends towards the source of the stimulus. This effect is called positive phototropism and it might be thought that roots display an opposite reaction, growing away from the light and into the dark of the soil. But even aerial roots grow downwards, and it has been shown that root growth is in fact a positive geotropism – a directed effect towards the pull of gravity. In reality, the situation in many plant organs is more complex than this, for both positive and negative geotropic and phototropic effects can occur in the same plant organ at different times. Other types of light-stimulated growth exist also: the differential opening and closing of flowers or even of leaves during daylight or dark, for instance, and the specific requirements for the presence or absence of light in order for some seeds to germinate. Commonly, there is a temperature as well as a light or gravity effect in these and other hormone-mediated growth responses – witness the frequent requirement for seeds to be given a hot or cold response before germination will occur.

Perhaps the most interesting and in some ways the most important plant growth effects are those associated with the control of flower development. Part of the challenge and satisfaction of gardening stems from the fact that plants flower at different times and by careful choice it is possible to have blooms every month of the year. Why should this be? The underlying natural logic is not too difficult to appreciate for it is important that each type of plant flowers at a time of year when its pollination mechanism is likely to be effective (in many instances this means when the appropriate insects are active) and when there is a long enough period for seeds to be set and dispersed before a climatic change (a drop in temperature for instance) makes further growth that season a biochemical impossibility. Some system is needed, however, to stimulate the tissues at the stem apex to develop into petals and stamens instead of leaves, and it has been known since the early years of this century that daylength rather than temperature influences this. The phenomenon is known as the plant's photoperiodic response. Some plants flower when days are long in the summer, some when days are short in the winter, and some almost all year round. The three groups are called long-day, short-day and neutral respectively, and not surprisingly these differences generally reflect the latitude of their natural homes – short-day plants like salvias coming from low latitudes near the Equator and long-day plants like spinach from higher latitudes closer to the Poles.

Long-day plants will actually continue to flower in continuous light and the way in which this affects hormone activity is imperfectly understood. More is known, however, of short-day plants, where a different mechanism must operate as it is clear that it is the long night rather than the short day itself that is the critical factor. If short-day plants are grown in conditions of short days but then artificially exposed to bright red light of short duration during the night, flower initiation is retarded. This phenomenon is used commercially to retard flowering in such short-day plants as chrysanthemums so that they can be produced in bloom at any time of the year. The chemical receptor of the stimulus is a protein called phytochrome and it exists in two forms, changing from one to the other with light of different wavelengths. One form seems to stimulate hormonal or other chemical changes to take place at the stem apex and so initiate the development of flowers. Interestingly, a leaf from a plant that has received a flowering stimulus of the appropriate daylength can be grafted onto one of a quite different species that has received no such effect and bring about flower development.

It will be apparent to anyone who has ever accorded plants more than a passing glance that, as they progress from seedling to senescence and ultimate death, they pass through changes that are more than simply ones of size. Leaf shape in particular often changes as plants mature and sometimes, as in species of *Eucalyptus*, the plant is deliberately cut back annually to remove the older shoots and so encourage the formation of the more attractive juvenile foliage.

The reasons for many of these differences are not known but a very obvious one that is certainly important in gardens is the change in leaf colour that occurs in many deciduous plants as the individual leaves reach the end of their lives and

One of the glories of autumn is brought about simply by the death of leaves. As photosynthesis ceases, chlorophyll production stops and other pigments take over, the dramatic colours of shrubs and trees like the maple become apparent.

days and cold nights that occurs in the American Fall. The actual dropping of the leaves of deciduous plants arises from the development of an abscission zone at the base of the leaf stalk, and within this zone the middle layer of the cell walls softens. This process, too, is under hormonal control, as evidenced by the fact that whilst leaves on a dead branch will shrivel, they do not drop. A classic, graphic and extremely useful example of the hormonal influence arises with the beech (beech hedges unlike beech trees do not shed their dead leaves in the autumn, rendering them valuable as screening plants). Clipping retains the shoots in a permanently juvenile state in which the auxin produced at the shoot apex is present in sufficient concentration to inhibit the enzymes that are needed to break down the cell wall. In the long, normal, unclipped shoots of a tree, the auxin is so diluted that abscission can proceed. Interestingly, too, some fungal pathogens that induce premature leaf fall on affected plants do so by inducing the production of abnormally low levels of auxin.

In fruit, senescence is usually called ripening and it, too, is characterised by specific chemical changes that have importance for gardeners – there is often a change in the storage carbohydrate from starch to sugars with a consequent increase in sweetness, whilst the chemistry of the cell walls changes to result in a general softening of the tissues. The ripening process (or, at least, the increased respiration that is part of it) is enhanced by minute quantities of the gas ethylene which is actually produced by plant tissues in appreciably greater amounts as the ripening itself progresses. Thus if a ripe fruit is placed among unripened ones, it will speed up their ripening in turn.

I hope that this chapter will have shown that, like other living things, plants are complex, both in their structure and in their dynamic functioning. This is a situation that should fascinate and stimulate gardeners. For in their complexity, plants challenge us, if not quite to master them, at least through understanding to influence their many inter-related processes and bend them to our will.

senesce in autumn. Chlorophyll production slowly ceases, thus exposing other coloured pigments in the leaf, especially yellow carotenoids. Sometimes, further pigments such as the red anthocyanins are produced at this time, thus enhancing further the very attractive appearance of autumn foliage. At least some of these processes are temperature as well as daylength induced, and the reason that so many North American maples and other trees fail to produce such spectacular colour in Britain can be attributed to the greater contrast between warm

CHAPTER THREE

The plant and its environment

This book is essentially about the cultivation of plants and the site within which they are grown, the garden. I think of this simply as an area adjoining a dwelling where plants are cultivated for the needs of the inhabitant, rather than for trade or sale. One of the biggest mistakes that anyone can make is to assume that when taken from its natural environment and habitat and cultivated in a garden or anywhere else a plant suddenly begins to obey different rules and respond to different influences. A cultivated area like a garden differs from a natural plant habitat solely in that certain environmental features have been suppressed and others enhanced at the hands of us, the gardeners. But by looking at natural plant habitats you will soon learn that the birth, growth, reproduction and death of any individual plant cannot be considered in isolation from a multitude of other features relating to other plants and to other organisms growing in the same habitat. So it is in a garden too; you cannot pull up a weed, plant a bulb, apply fertiliser to your lawn or water your brassicas without also affecting several quite different aspects of the lives of the plants concerned and without affecting other components of your garden.

In this chapter, therefore, I shall consider the various types of environmental factor that govern the ways in which plants grow and that we as gardeners make use of in our horticultural endeavours. And I shall do so by using the three main categories that ecologists use, for ecology is the study of habitats and the interactions of organisms. Ecologists recognise the main environmental influences that affect plants as climate, soil and other living organisms.

Climate is perhaps the most talked about environmental factor and almost always the first

subject mentioned when two gardeners meet. As far as Britain and most other temperate countries are concerned, it is the general unpredictability of the weather that provides the principal conversation piece, for of course the various basic physical features that make up climate and weather are the same the world over. Long- and short-term maximum, minimum and average temperatures, solar radiation (including, of course, light), precipitation and wind are the climatic features that affect all plant habitats, natural or artificial.

All plants, and to a certain extent all other living organisms, rely on rather similar basic biochemical processes in order to grow and, like all chemical reactions, these are either speeded up or slowed down by particular temperatures. For this reason, most plants actually grow best and quickest within the same overall temperature range – they grow little above 25 °C or below 5 °C and best between about 10 and 20 °C. By and large, therefore, tropical forest plants grow more quickly than temperate climate species because the optimum temperatures for growth occur for a much greater proportion of the year. And by the same token, all plants that occur in extremely hot or extremely cold conditions (cacti in deserts or alpines in polar regions), tend to be slow growing.

Nonetheless, whilst the length of the period during which the optimum growth temperatures occur may dictate how successfully a particular plant will thrive, it does not govern the basic matter of whether it will grow at all in any given place and it is another temperature measurement that has the greatest influence here. This is the annual minimum temperature or, more specifically, in temperate climates, the average monthly minimum temperature. (However, this measure should also

A garden comprises a range of different habitats, varying in soil type, relative exposure to weather, and other factors. It is part of a gardener's skill to choose plants appropriate to each particular spot, by appreciating the conditions under which the various species grow naturally.

be considered in combination with a plant's possession of those protective features that give it the ability to survive when temperatures fall to potentially damaging levels.) The difference between the average annual minimum and the average monthly minimum temperature is best demonstrated by reference to a specific example. The average monthly minimum at St Mary's in the Isles of Scilly (49°56′N 6°18′W) is 5.6 °C which suggests that frosts never occur. But in fact the average annual minimum is -0.6 °C and close examination of the average January and February minima (both 1.1 °C) indicates that at least slight winter frosts must be fairly common.

Almost all energy on earth originates with the sun and with the solar radiation that reaches us through space. More than half of the total radiation that strikes the earth's atmosphere is reflected back into space or absorbed by the atmosphere, but we manage fairly well with the remainder that penetrates to the surface. Even so, the actual amount reaching any particular site will vary locally with such factors as the density of the cloud cover (which increases the reflection back into space), with the time of year and with the geographical situation, the last two both dictating the angle at which the sun's rays strike. Gardeners will probably be familiar with the concept of expressing radiation in terms of wavelength through their knowledge of radio and television waves but all forms of radiation can be described in these terms too and the variation in wavelengths of solar radiation are immensely important for plant life. The total radiation spectrum ranges from radio waves with wavelengths measured in hundreds or even thousands of metres to cosmic rays with wavelengths of less than a nanometre (a thousand millionth of a metre), but the two most important components for plants are visible light (needed for photosynthesis and for the control of flower initiation), which ranges in wavelength from about 380 to 780 nanometres, and the longer wavelength warming radiations tending towards the infra-red (800 nanometres to 1 millimetre), needed to provide the temperatures required for growth. (Radiations of wavelengths of about 10 to 400 nanometres are called ultra-violet and are of interest to gardeners because they can degrade some of the plastics used in gardens, as well as being responsible for gardeners acquiring sun-tans when the work is finished!)

The part of the visible light radiation spectrum actually used by plants for photosynthesis is called photosynthetically active radiation and falls within the wavelengths from 400 to 700 nanometres. Although it was once thought that the spectral

Although much active plant growth ceases in autumn and winter, gardening interest should not necessarily cease also. A careful choice of subjects ensures that a garden's visual appeal continues throughout the year.

Careful studies, such as those being performed in these experimental glasshouses, have revealed that the virtual monochromatism (single wavelength) of orange sodium light is surprisingly the most valuable for plant growth.

pattern of radiation (in effect, the proportion of light of different colours) within these wavelengths was critically important this is now known not to be so. And nowhere is this better demonstrated than in the artificial lighting used for commercial greenhouses. Whilst growers formerly used lamps of a type emitting radiation that as closely as possible mimicked natural daylight, most new commercial installations have low-pressure sodium discharge lamps which produce virtually monochromatic light with a wavelength of about 589

nanometres (similar to the orange light familiar from street lighting systems). It is therefore the total amount of radiation within the photosynthetically active band that is important, rather than the relative amounts of red, green, blue and so forth.

Because the total amount of radiation (including both photosynthetically active visible light and warming infra-red) reaching the earth in temperate latitudes is much less in winter than in summer, plants may not grow in such regions satisfactorily all year round. Local effects which tend to enhance

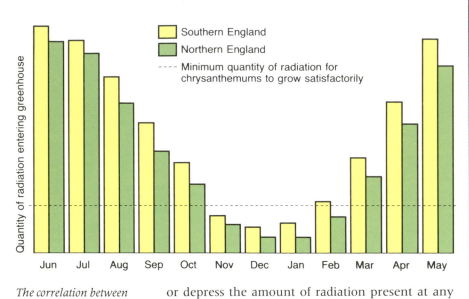

Quantity of radiation entering greenhouse

- Southern England
- Northern England
- ---- Minimum quantity of radiation for chrysanthemums to grow satisfactorily

Jun Jul Aug Sep Oct Nov Dec Jan Feb Mar Apr May

The correlation between photosynthetically active radiation entering greenhouses in northern and southern England and the satisfactory growth of chrysanthemums. This indicates clearly when commercial growers require and benefit from the use of artificial lighting.

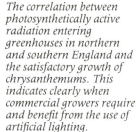

The almost total inability of any plants to grow at temperatures below freezing does not preclude their ability to tolerate it. Holly leaves can survive unharmed at a temperature of −16 °C, which would kill many other species.

or depress the amount of radiation present at any particular place are generally of more significance in relation to infra-red and temperature than they are to visible light and illumination. This is because there is no critical cut-off point for visible light, below which growth fails. With infra-red radiation, however, there is such a cut-off; it is the point at which the amount of warming radiation becomes inadequate to prevent freezing. Any small factor which tends to tip the temperature downwards below freezing can be critical, and whilst hardy plants are usually well enough equipped to withstand this in the depths of winter (this ability being, in effect, a definition of hardiness), the occurrence of a late frost in spring after growth has commenced can be devastating.

Most winter frosts arise when cold air is transported over the ground from one site to another, and low-lying places where this dense cold air collects are called frost hollows. Most spring frosts, however, are radiation frosts; they arise when the heat loss from the soil during the night is greater than is compensated for by the day-time warming up. A clear sky in spring before the day-time sun provides very great warmth can create just such conditions. The date at which the last spring frost

occurs varies mainly with the length of the night-time cooling period. The map on the right shows how this is manifested in different parts of the British Isles, but spring frosts are also more likely where there is a dry, free-draining soil – this loses heat rapidly at the surface and is a poor conductor of heat from reserves at depth. Ironically, spring frosts are also frequent where there is a good cover of vegetation for, whilst the vegetation keeps the soil warmer, it causes the air immediately above it to cool down. Blossom trees growing among grass are, therefore, more prone to spring frost damage than those in bare soil. And remember that the important frost for plant life is the air frost – one where the temperature is 0 °C or less at a height of 120 centimetres; by contrast, a ground frost, which occurs when the temperature has fallen to 0 °C at or just above ground level, is of much less importance, for the basal parts of plants are usually better protected.

Whilst it is fairly apparent that various plants have adapted to the differing levels of temperature and light in different parts of the world, there is also a wide range in preferences even among the plants of a small area. If you stand on the floor of a woodland, you will soon note that it is both cooler

SCALE

1 15 1 15 1 15 1 15
JUNE MAY APR. MAR. FEB.

The Siberian larch might be thought able to withstand the relatively mild British winter, but it can be stunted by frosts that may occur unexpectedly after a few days of mild weather.

The average dates in the British Isles of the last spring air frost; an indication of the time of year when half-hardy plants may safely be placed outdoors.

and darker than outside in a field and that the plant life is very different; there will be very few grasses, for instance, in the wood and very few bluebells, dog's mercury or wood anemones outside. These are observations worth remembering when you come to choose plants for particular places in your garden.

The second of the important climatic features for plant life is precipitation, a useful word to cover all of the forms in which water falls to earth from the atmosphere (rain, snow, hail, drizzle, sleet, hoar frost, fog and dew). At least some of the importance of water to a plant will be apparent from Chapter Two. It will also be apparent that a growing plant's relationship with water is not static for water passes through it continuously and only when a deciduous plant is dormant and leafless is this through-flow seriously stayed. As is well known, moreover, water itself moves cyclically through the environment as a whole, falling as rain or other precipitation from clouds, most of it then draining through the soil to rivers and the sea (and some being intercepted by plants *en route*), to return into the atmosphere by evaporation. It is obvious that even within a small area, rainfall can vary markedly, but in the British Isles as a whole, the south and east are clearly drier than the north and west. This is because in general the nearness of large ocean masses and the presence of high land mean higher rainfall – warm air passing over the ocean collects moisture by evaporation, then rises and cools as it reaches higher ground and the moisture condenses and falls as rain. Rainfall also occurs in low-pressure systems when moist air blown from the south meets cold polar air blown from the north and rises above it. It is largely rainfall of this type that falls over low-lying areas but, because, unlike mountains, masses of cold air are constantly moving, it is much harder to forecast. Thunder rain is harder still to predict, arising when localised heating of the earth causes powerful upcurrents of warm, moist air within which large raindrops form and subsequently cause appreciable damage to plant life when they fall.

Snow is important for plant life, but not for its contribution of moisture as it almost always falls when plant life is dormant and in any event contributes only one-tenth the amount of water of a comparable depth of rain. But snow offers invaluable protection against penetrating frost; in most winters, a covering of 7.5 centimetres of snow will adequately prevent plant life beneath from freezing. A very cold, very dry winter, therefore, without a protective snowy blanket (such as that of 1986–7 in many parts of Britain) is one when winter frosts can be particularly damaging to plant life. Conversely, a large bulk of snow lying on the branches of trees and shrubs can cause breakage and serious damage, especially to evergreens which present a large surface area on which the snow can accumulate. The narrow shape and slippery surface of conifer needles combined with the overall pyramidal tree form of northern species is often thought to be an adaptation to facilitate the sliding off of snow.

Lack of water is obviously important to every individual plant, but although some parts of Britain experience periods of drought in most years, it should be remembered that there is no area where shortage of rain prevents a particular native plant species from growing if other environmental factors are suitable. Even after the prolonged drought of 1976, when much of the British countryside was parched and brown, native vegetation was lush and green again within days of rain recommencing. Nonetheless, some garden plants originate from naturally wetter regions of the world and are thus less able to cope with water shortage. And, even more importantly, many garden plants have been selectively bred with features that render them abnormally prone to drought (vegetables with very large leaves, for instance) and will inevitably not give of their best when denied water for more than short periods.

The least appreciated of the important climatic features that affect plant life is wind, the natural movement of air. Only those gardeners who live on the coast or in the hills are reminded constantly of

Whilst snow may form an insulating blanket to protect plants against penetrating frost, its sheer weight can result in damage to branches, especially of evergreen species.

its influence. But wind affects plant life in several ways. It is a powerful physical force – for instance, a wind of speed 88–102 kph (55–63 mph) which is called a Whole Gale, is described on the Beaufort scale of wind speed as characteristically causing the uprooting of trees.

Wind is also a very important medium for the transport of pollen from one flower to another (see p. 116), for dispersing fruits and seeds away from parent plants to enable particular species to colonise new land and for carrying potentially harmful insects and fungal spores which may have consequences for plants on which they land. Salt spray from the sea and, in more recent times, noxious chemicals (including those responsible for causing acid rain) are also carried by the wind.

However, it is perhaps the drying effect of wind that is of greatest overall importance for plant life. Only because wind brings about the evaporation of water from leaf surfaces can the transpiration pull operate, draw up moisture from below and maintain a plant in a viable state. But enough wind can very soon become too much – the rate of drying at the leaf surface can easily exceed the rate at which the roots are able to take up moisture to replace that lost. There are several ways that this can affect plant growth. Very strong, drying winds cause buds to shrivel and thus limit leaf growth on the side of the plant that faces the wind. This is called wind pruning and its effect is to produce the distorted trees which seem permanently to be leaning away from the wind, seen on the sea coast or other exposed areas. During periods of summer drought, the amount of water in the soil is simply inadequate to replace that lost through the leaves and plants wilt in consequence. But in winter, too, evergreen plants continue to lose water by evaporation (although most do have leaves structured to limit this loss, see p. 21), and whilst the soil may be full of water, it is unavailable to the roots if it is frozen solid. Up to a point, plants can tolerate these water shortages yet recover when supplies are resumed. But beyond that point, damage is brought about as the protoplasm inside cells is drawn away from the

The drying effect of the wind is particularly apparent in those plants growing in exposed locations which appear permanently to be leaning. This is brought about when the buds on the windward side dry out and are killed.

cell walls; this damage is irreparable and the leaf or entire plant will then die. I shall describe in Chapter Four the impact of this in gardens and how it may be countered. Nonetheless, even when plants can tolerate the drying effect, growth may be stunted; compare individuals of a plant species growing on a high exposed mountainside with those in a sheltered valley bottom and they may be barely recognisable as the same. The importance of shelter in a garden in enabling you to grow plants worthy of their name is one that I believe in passionately and later (see Chapter Nine) I shall describe the best ways to provide this.

A dedicated naturalist or gardener will find interest and reward in making his or her own weather measurements or, even more pertinently, of amplifying official local weather information with measurements of the small variations that constitute the microclimate within a small area. But even if you do not aspire to such enthusiasm, much of importance about the microclimate can be deduced simply from an examination of aspect. Does the site face north or south, is it shaded by trees or buildings? Does it lie close to a river bank or at the top or foot of a slope? Merely by using your eyes and your common sense, you will learn a good deal about the climatic reasons why certain plants are growing naturally in any one spot or the likelihood of particular species succeeding in a garden.

Almost all plants grow in soil, but whilst we take for granted the fact that wherever we happen to be, soil very probably lies somewhere beneath our feet, the immensely variable nature of this wonderful material is barely appreciated. Even to talk of *the* soil as if it was but one substance is most misleading. It is a constantly changing medium, a microcosm of life on earth. It is through the soil that plants obtain almost all of their water requirements and their mineral nutrients, and it is the soil that holds them upright. Yet it is also the source of many of their problems in the shape of other plants that compete with them and of fungi, bacteria and insect and other pests that would cause them ill. It varies in countless ways, but for plant life the most important variables are its overall chemistry, its texture and its structure. And in thinking about the factors that contribute to these features, it bears remembering that all soils contain differing proportions of three main types of substance – inanimate matter that has never lived, living matter and matter once-living but now dead.

The inanimate matter comprises the soil minerals together with the soil water and the soil air. Soil minerals are derived from rocks that have been eroded and degraded over thousands of years by rain, rivers, wind, frost, glaciers and the sea. These same forces can move the degraded matter many miles from the position of its parent rocks, so whilst the geology of a region does often give a good indication of the likely chemistry of its soil, the minerals present in a soil made up for instance of river alluvium or glacial debris may be very different from those in the local underlying rocks. The degradation of mineral matter from rocks results in three main types of particle: sand (with particles 0.06–2.0 millimetres in diameter), silt (with particles 0.002–0.06 millimetres) and clay (with particles less than 0.002 millimetres). The relative proportions of these three types of particle give a soil its texture. The texture triangle above right shows the names given to soils on this basis.

However, inanimate matter alone does not make a soil, nor is it the only feature contributing to soil chemistry. For once minerals accumulate in any particular site, they will be colonised by plant life, at first usually of the relatively simple forms such as mosses. When these plants die, their remains become added to the soil to constitute the first of what I have called the once-living but now dead component known as humus or simply as organic matter. Gradually more advanced flowering plants begin to grow too, and these in turn add more humus, so changing the overall chemistry of the soil and also influencing the degree to which the loose mineral particles adhere together by natural glues to form crumbs. These processes of soil formation can sometimes be seen, apparently frozen in time, if a deep hole is dug, especially into an area where the soil has been little disturbed by cultivation. More or less distinct layers or horizons will be visible, ranging from parent bedrock at the base,

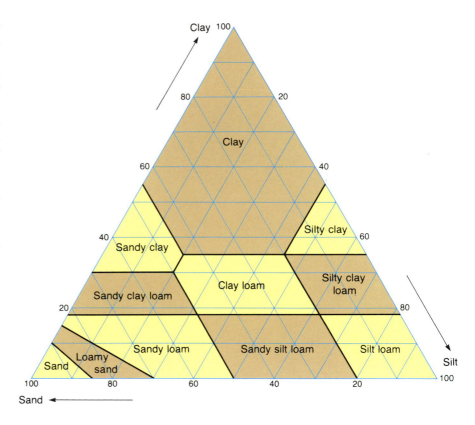

The method used by the Soil Survey of England and Wales to name the major soil groups according to their relative percentage contents of sand, silt and clay. The individual soil types found in particular areas are further designated under categories known as 'associations'.

A soil profile revealed when a hole is dug vertically into the soil and subdivided into the main layers or horizons. Such a profile is only likely to be seen in soil that has remained relatively undisturbed through the activities of cultivation.

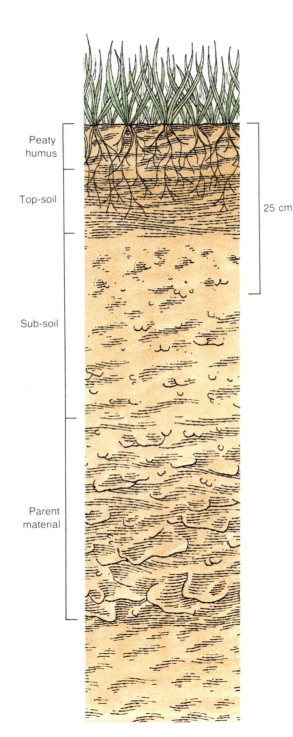

Peaty humus

Top-soil

25 cm

Sub-soil

Parent material

The most important decomposers of organic matter to form soil humus are fungi. The above-ground appearance of toadstools belies the many metres of microscopic threads of mould that exist below the surface.

through the sub-soil (a region of slightly changed and eroded mineral matter largely lacking any organic component), to zones above with increasing proportions of plant and animal remains – the top soil. The relative depth and numbers of these horizons are characteristic of particular types or, as they are called, series of soils.

The degree of aggregation into crumbs and, in consequence, the proportion of the soil that comprises the solid matter and the proportion that comprises the spaces or pores, gives a soil its structure. A soil with a blend of large pores between the crumbs and small pores within the crumbs is called well-structured. Loams are examples of such soils. A soil that comprises a large proportion of either clay or sand, having few crumbs and with, in a clay, almost no pores at all or, in a sand, with very large pores between the particles only, is said to be almost structureless.

The pores within and between soil crumbs are not simply voids. They are filled either with water or with air and, consequently, a soil that is highly aerated cannot at the same time be waterlogged. Water is also present adhering to the surface of the crumbs and the entire water content of the soil is called the soil water reservoir. When a particular soil is holding the maximum amount of water possible in conditions that permit free drainage, the soil is said to be at field capacity. To gain an idea of what this means, envisage a sieve filled with soil onto which a large volume of water is poured. At first, water will drain freely through the sieve but eventually will stop doing so. (Roughly speaking, water drains under the force of gravity from pores that are more than about 30 micrometres in diameter but is retained in smaller ones from which it can, therefore, only be removed by a force that exceeds gravity.) The soil is now at field capacity and the amount of water retained in the soil varies greatly, depending on the size and number of the pores which in turn depend on the two features I have already mentioned, its texture and its structure. As plants remove water from the soil or as it is dried directly by evaporation in the height of summer, and no additional water is supplied, clearly the water content will fall below field capacity. The amount that it falls is called the soil moisture deficit and this is usually expressed as the number of centimetres of rain or irrigation needed to restore field capacity. This is an important measure and in due course I shall be discussing its relevance to the amount of water you should apply to your garden. Although, of course, soils vary widely, a good cultivated loam at field capacity should have about 50 per cent of its total volume taken up by pores; however, only about 85–90 per cent of this pore space (about 45 per cent of the total soil volume) is filled with water, the remainder being air.

The drier a soil becomes, the harder it is for plants to remove water at all, for that remaining tends to be held increasingly tightly in the smallest pores and on the soil particles. Eventually, the force with which the water is held becomes so tight that the 10 atmospheres or so of pull that roots exert is quite unable to extract sufficient to keep the plant growing. The plant loses its turgor (see p. 13) and begins to wilt. The soil moisture can then be said to be at wilting point and the difference between the amount of water in a soil at wilting point and a soil at field capacity is called the available water. In a poorly structured clay or sand, the amount of available water is small – in the case of clay because both the field capacity and wilting point are very high and in the case of sand because they are both very low. Neither sandy nor clayey soils will therefore enable plants to thrive for long in the absence of rain before they wilt. It is perhaps worth saying that plants do not atually die once the wilting point is reached for roots can still take in a little water even when a pull of up to thirty times atmospheric pressure is needed. But that really is the limit, and beyond this point the plant will be irreparably damaged even though some water does still remain in the soil. Indeed, to remove the last vestiges of soil water, a suction pressure of some 10 000 atmospheres would be needed; this is the pressure required to support a column of mercury 7.6 kilometres high.

When mineral matter is outweighed in any particular soil by humus, the soil is generally called an organic soil. By virtue of their structure and chemistry, such soils are often excellent supporters of plant life, but one of the most extreme organic soils is a peat and this is almost bereft of any plant nutrient. Peat is formed entirely from partly decomposed plant remains; partly decomposed because it forms on sites where because of impeded drainage (brought about for instance by a dense clay beneath) water accumulates, air is excluded and the bacteria and fungi responsible for decomposition cannot thrive.

In mentioning peat, I have introduced the two most important aspects of soil chemistry for plant life – the amount of plant nutrient the soil contains and its relative acidity or alkalinity, known as the pH. Almost all of the plant nutrient in a soil derives from its mineral matter and is thus a feature of the type of rock from which the soil is formed. I shall be describing the importance of different soil chemicals for plant growth in Chapter Five, but it must be said that no British soils are naturally so deficient in any of them that plants cannot grow. Certainly, different types of plant may thrive better on some soils than others because of the differing proportions of nutrients present, but on any particular soil the substances removed by plants are generally replaced when those plants die and decompose and the situation overall remains stable. Gardens, nonetheless, are very different for there is often no balance between nutrient removed and nutrient replaced, as will be seen in due course. But if actual nutrient content is not a major factor in limiting plant growth on any particular soil, pH certainly is, and although the term has entered common gardening parlance, it is still not widely understood. It stands for 'potential of Hydrogen', actually a rather complex chemical measure of the concentration of electrically charged atoms or ions of hydrogen, and it is expressed on a logarithmic scale – in other words, a pH of 6 indicates ten times as many hydrogen ions as pH 7 and one hundred times as many as pH 8. The number of hydrogen ions gives an indication of what is usually called relative acidity or alkalinity and the entire pH scale runs from 0 (extremely acid) to 14 (extremely alkaline). The mid-point of the scale, pH 7, is called neutral. But these extremes are solely of interest to a chemist (concentrated hydrochloric acid for instance has a pH of about 1), and soil pH covers a much narrower spectrum and ranges in Britain from about 3.5 for a *Sphagnum* peat to about 8 for a thin soil overlying chalk. Soils are named in relation to their pH as: strongly acid (below 4.5), moderately acid (4.5–5.5), slightly acid (5.6–6.5), neutral (6.6–7.5) and alkaline (above 7.5).

There are three main and related reasons why pH affects the growth of plants. The first is their actual tolerance of high or low concentrations of hydrogen ions. This is probably the least important practically. The second is their need for and/or tolerance of high or low levels of calcium because, in Britain at least, an alkaline soil usually contains a large amount of calcium (chalk and limestone are forms of calcium carbonate) whereas an acid soil contains little. Plants highly tolerant or intolerant of calcium are called calcicole and calcifuge respectively. Many plants have no strong reaction either way, but it is often the occurrence of wild plants with calcicole or calcifuge features that gives the natural vegetation of an area its characteristic appearance (clematis on alkaline and heathers on acid soils, for instance) and observing the natural vegetation of an area can be a useful first guide to the type of soil occurring there.

The third reason why soil pH affects plants is related to the nutrient content. I mentioned earlier that natural soils with an *actual* nutrient deficiency are so rare as to be discounted, but in alkaline conditions most nutrients combine with other soil chemicals in such a way as to render it difficult for plants to take them up (see pp. 66–8). The nutrients are said to be unavailable. And whilst plants such as clematis and heathers have adapted to enable them to thrive in more extreme conditions, on balance the optimum soil pH for most plants to acquire most nutrients is about 6.5.

I shall discuss the remaining inanimate constituents of soil (the soil water and soil air) in some detail in Chapter Four and shall therefore now move on to the third of the three major types of soil component, the living matter. This is at the same time extremely varied, extremely large and rather inadequately understood. In addition to plant roots, the soil is home to forms of almost all types of lower organism – algae, mosses, lichens, fungi, bacteria and protozoans, for instance, and almost all major groups of animals have species that spend at least part of their lives in the soil. Earthworms, eelworms and many insects, mites and other arthropods are obvious examples but some mammals (rabbits, moles, voles), reptiles (lizards, snakes), amphibians (toads) and even a few birds like the burrow-nesting martins occupy this environment for a greater or lesser proportion of their time. The importance of soil-inhabiting organisms, whatever their size, falls into three main categories. The microscopic fungi and bacteria, together with some of the animals like earthworms, are directly responsible for altering chemically the soil components by the activities of their digestive enzymes. Thus they

Heaths and heathers (and, indeed the family Ericaceae in general) are the plants most typical of highly acidic soils in temperate climates, although a few heather species are fairly lime-tolerant.

convert the rather complex substances that make up the dead remains of plants and other animals into the simpler forms that living plants can utilise as nutrient. And some specialised species of bacteria and fungi have other very important roles to play in plant nutrition through the functioning, respectively, of root nodules (which convert atmospheric nitrogen into inorganic salts) and mycorrhizas (which assist roots in the uptake of nutrients).

The second way that soil life is important is in moving, mixing or simply stirring up the soil to produce a blend of the various ingredients. The action of moles in bringing large quantities of soil (and weed seeds) to the surface is very evident, but the action of smaller creatures is actually much more significant. Earthworms are especially important and it has been calculated that 100 tonnes of

soil per hectare are moved by the earthworm population in the course of a year. And during these soil movements, the worms form burrows which aid significantly both drainage and aeration and provide channels for encouraging root growth. But whilst earthworm populations are large (perhaps up to 1½ million per hectare), they are actually well outnumbered by arthropods and nematodes (eelworms). Estimates have been made of well over one thousand million arthropods per hectare. Mites are often the most numerous among them, followed by insects and, especially, collembolans (springtails). This seems vast enough, but the nematode population under a hectare of soil surface can actually exceed one hundred thousand million. However, the size and bulk of an earthworm, coupled with its way of life, make it an extremely efficient item of earth-moving equipment, as important in a garden as in a natural habitat.

The third way that soil organisms affect plant growth is through their roles as pests and as the causes of diseases. Insects and nematodes are especially important in the former role and fungi in the latter. I shall be discussing this subject in detail in Chapter Eight but it should be borne in mind that other members of the soil flora and fauna have their own parts to play in keeping such potentially harmful organisms in check. The soil, like the above-ground plant environment, is a manifestation of what we call the balance of nature.

So the plants that grow in any particular area do so primarily because of a special combination of climatic and soil features. But it is very rarely that only one plant species finds this combination to its liking, and there will always be many animal species attracted to a site, very largely by the plants themselves. So plants live naturally not as isolated individuals but in communities in which all the species interact. Such interactions constitute much of the stock in trade of ecology, a quite extraordinarily fascinating subject that has been greatly ignored and unappreciated by gardeners in the past. I can do no more here than introduce the two ecological notions that I feel are most pertinent to anyone who gardens and hope that in doing so I shall encourage you to look anew at natural and semi-natural plant habitats, such as woods, hedgerows, meadows and cliff-tops, and then try to envisage similar effects operating in your garden.

Most people have encountered the idea of a food chain – cows eat grass, people eat cows, being a very simple version. In practice, the interactions in any particular habitat are more complex than this. They are also multidirectional and are better described as webs rather than chains. I think that people are also now used to thinking of food in terms of energy – look at any packet or can of food from your supermarket and you will see an energy value quoted on the label. (The energy value of food is actually expressed in the form of kilocalories

(a measure of heat) or kilojoules (a measure of work).) Try to think of gardens and other habitats in energy terms too, and realise that all energy comes originally from the sun, is converted into chemical form by plants and then moves into and out of the habitat in some of the ways that I have shown opposite. Anything that interrupts one form of energy input or output will have far-reaching consequences therefore – think of the effects on the habitat shown opposite if the tree is cut down.

The second ecological feature of a habitat to consider is summed up by the three words, invasion, succession and competition. Take any area of bare land – a rock outcrop, a sand-dune, a lake edge, for instance, and imagine spores and seeds gradually being blown or carried on to it. These will germinate and, if the conditions are suitable for any particular species, they will grow and begin to change the habitat slightly as they help soil to form, as I have outlined on p. 39. They will also change the microclimate slightly and alter the habitat to create conditions conducive to other plant species which will invade and colonise too and compete with the originals for light, air, nutrients and water. Gradually, the invaders may take over and thus create a succession. In turn, they too may be ousted until, for every natural environment, a more or less stable community of plants arises within which individual plants grow and die but where the overall structure, appearance and species composition remains much the same. This is called a climax community and, in general, it is the most complex type of community and includes the largest species that the local soil and climatic conditions allow. Over much of Britain, the climax community is a deciduous woodland. The reason I am explaining this is because your garden was once a natural community, probably a woodland. It is not frozen in time but is maintained in its present form solely by your efforts. Relax them, allow weeds to establish, and your garden is on its way to becoming woodland once again.

Before I move on to discuss the principles behind some of the more practical aspects of gardening, it is

A diagrammatic representation of the flow of energy and matter into and out of a typical habitat, together with the most important living things representing its more or less constant inhabitants. Such a system can be modified to fit all natural and semi-natural environments.

INPUT
Solar energy

OUTPUT
Heat losses

Energy

INPUT
Animal
immigration

Primary
producers
plants

OUTPUT
Animal
emigration

Matter

INPUT
Rain

Herbivores

Leaching

Energy +
matter

INPUT
Dust

Carnivores

OUTPUT
Cropping

Biomass

Faeces
litterfall
etc.

UPTAKE

Litter
breakdown

ROOT
Breakdown

INPUT
Soil &
rock
weathering

Decomposers

OUTPUT
Drainage

appropriate to introduce one further feature of the way that a plant is equipped for life in any particular habitat. This is important for it reflects the manner in which even genetically unrelated plants have adapted to the constraints of their environment and it can tell a gardener a great deal both about their natural habitats and about the way they might fit into a planting scheme.

The individual cells and tissues of a plant comprise its physical structure or anatomy. But the way that these cells and tissues combine together produces something of unique form or morphology. Although plants are named and classified in a way that tries to reflect their genetic relatedness, it is apparent that there is another way that they can be grouped; a way that makes use of their overall morphology. Whilst the ability to put peas, lupins and the Mount Etna broom into one and the same family requires a degree of knowledge about the importance of flower structure (see pp. 22–3), even a child with no biological training at all would easily group together oaks, pines, elms and mahogany into a category different from plantains, dandelions and daisies. Such a grouping is usually called a life form classification and the system used most widely today groups plants according to the position relative to soil level occupied by the buds that survive during periods of dormancy to recommence growth in the following season. This system divides biennials and perennials (plants that live for two or for more than two years respectively) into four main types together with a group comprising annual plants that have no survival buds, for each individual lasts for one season only. I have listed the various categories in the table opposite together with familiar garden examples of each type, and below offer examples of the ways that particular life forms benefit plants for particular types of habitat. The rosette plant, like the dandelion, with its buds very close to soil level, is ideally equipped for a habitat populated by grazing animals whose browsing and trampling would cause considerable damage to types with aerial buds. It also equips it ideally for life with the lawnmower. The bushy,

creeping, flopping and cushion life forms on the other hand are excellent for survival in cold climates such as Northern Norway. The winter snow protects the buds close to soil level but they are well placed to begin growth quickly in a short summer when the air near the ground warms up first. Annuals tend to be opportunist plants and make use of any environment and climate where the growing season is short – in a temperate climate summer or in the brief aftermath of rain in a desert, for instance. So always bear in mind that the appearance of a plant is not some accident of nature or haphazard assembly of the constituent parts. It can actually tell you something interesting and significant about the way that the plant lives and the conditions that it needs.

And lastly, before leaving ecological matters, a question that puzzles botanists and gardeners alike. Why should some plants, the deciduous types, shed their leaves more or less simultaneously whereas others, the evergreens, do so piecemeal and in consequence have foliage all year round. Certainly, the difference in impact on a community of having evergreens rather than deciduous trees is considerable – the evergreen casts shade all year round and is much more prone to be damaged by snow in winter, but can begin growth more rapidly in spring without waiting several weeks for buds to swell and burst. Among major groups of plants, the conifers are almost all evergreen and there tends to be consistency within families or genera of flowering plants too, although there are several examples like *Magnolia*, *Berberis* and *Lonicera* that include both evergreen and deciduous species. Nonetheless, there seems no botanical basis for the difference although, in terms of evolution, the deciduous habit is the more recent development.

Life forms of plants

Life form	Examples among garden plants
1. Survival buds borne on shoots well above ground level	
a. Trees and shrubs (evergreen or deciduous)	
Large trees more than 30 metres tall	Silver fir, hemlock, horse chestnut
Medium trees – 8–30 metres tall	Service tree, magnolia, willow
Small trees – 2–8 metres tall	Apple, medlar, juniper
Shrubs – less than 2 metres tall	Rose, potentilla, weigela
b. Epiphytes (plants growing on other plants, especially on trees)	Some orchids, bromeliads, ferns
c. Stem succulents (plants with large, fleshy stems)	Cacti, some senecios, euphorbias
2. Survival buds very close to ground level	
a. Bushy plants (aerial shoots die away to leave buds at stem base)	Michaelmas daisy, border phlox, solidago
b. Flopping plants (aerial shoots flop over so that buds are protected close to the ground)	Some clematis, stellarias and labiates
c. Creeping plants (aerial shoots creep and root at ground level)	Many thymes and campanulas, creeping Jenny
d. Cushion plants (very compact creeping plants, confined to a cushion form)	Thrift, many saxifrages, some artemisias
3. Survival buds at ground level, all aerial parts die away	
a. Basal bud plants (survival buds arise at ground level)	Raspberry, ceratostigma, gypsophila
b. Partial rosette plants (most leaves are in a basal rosette, but a few form on the aerial stem)	Chamomile, bugle, carnation
c. True rosette plants (all leaves are present in a rosette at the base of the aerial stem)	Some saxifrages, dandelion, daisy
4. Survival buds form below ground level or under water	
a. Earth plants (survival buds present on a rhizome, bulb, corm or tuber)	Iris, crocus, dahlia
b. Mud plants (survival buds are submerged in aquatic mud but aerial shoots extend above water level)	Water plantain, reedmace, arrowhead
c. Water plants (survival buds are borne below water and aerial shoots do not extend above water level)	Water lily, Canadian pondweed, hornwort
5. No survival buds; survival as seeds	
a. Annuals	Lobelia, calendula, radish

Percentages of different plant life forms among the total species of various climatic regions

	British Isles	Northern Norway (Arctic)	Mauritius	Nevada Desert
Large and medium trees	1	0	10	0
Small trees	3	0	23	2
Shrubs	3	1	24	21
Epiphytes	0	0	3	0
Stem succulents	0	0	1	4
Bushy, flopping, creeping and cushion plants	4	23	6	9
Basal bud and rosette plants	52	58	12	16
Earth plants	10	12	3	2
Mud and water plants	9	3	2	3
Annuals	18	3	16	43

CHAPTER FOUR

Cultivation and soil management

An expression that has passed in recent years into the everyday currency of the English language is 'The answer lies in the soil'. And whilst it may have originated as part of a comic act, there is nothing funny about the way that plants in one garden perform so much better than those in another because of the differing ways that two gardeners have treated and managed their soil. In the last chapter, I described the composition of soil and the way that the various components and features of it affect plant life. But garden soil is subject to several peculiar influences – the plants growing in it are usually alien to the habitat, they are very often removed before their naturally allotted life span, they are grown in artificial spatial arrangements both to neighbouring plants of the same type and to other types of plant nearby, and the expectations that the gardener has of them are greater than they might realise naturally. As a consequence of these influences, the soil must suffer a number of physical indignities, which themselves are interrelated as gardeners use them to rectify particular problems. For instance, soil is regularly subject to the impact of more water than falls naturally from the sky, it is forced to bear the repeated weight of gardeners' boots, lawnmower rollers and cultivator tines, and it is dosed at periodic intervals with alien chemicals. The soil in your garden is thus caught in a 'Catch 22' web. In Chapter Five I shall concentrate on chemical aspects of soil management, on plant foods and plant feeding, but first, here, I will deal with the soil's physical well-being.

The most important effect of the physical indignities that I have mentioned is that the structure of the soil is damaged. A heavy weight of any description on the surface of soil will compress together

The physical well-being of the soil and the extent to which it is managed by the gardener is reflected in the vigour of the plant growth it sustains.

the soil crumbs, especially at the surface, with consequent loss of some of the pore space and air content. Thus plant roots will be denied the air that they need and the lack of air space will, in turn, lead to impedance of the through-flow of water (and any nutrients dissolved in it) and also limit the growth of air-requiring or aerobic bacteria and fungi responsible for the degradation of organic matter. It might even, in severe cases, encourage the growth of anaerobic micro-organisms. These bring about a different degradation of organic matter that results in the formation of an amorphous dark material and the production of such gases as hydrogen sulphide, hydrogen and methane. Denitrification (the loss or removal of nitrogen or nitrogen compounds) is faster when the oxygen supply is impaired and thus the result is an accumulation of chemicals that may not only be useless as nutrients but may actually be toxic or otherwise detrimental to plant growth. The inability of water to drain freely through a soil will inevitably lead to its accumulation at the surface. This will further restrict the penetration of air, further aid the disintegration of the soil crumbs and thus result in a more permanent blocking of the pore spaces in the process called capping – especially serious on soils with a high content of clay minerals which pack closely together in a particularly impenetrable manner.

As the physical weight at the surface continues, its effect spreads downwards, and soil structure is impaired at ever increasing depth. Nonetheless, even a soil with a very high clay content and trampled by a small army of allotment holders is unlikely to be converted into a totally brick-like mass for several centimetres below the surface. But breakdown of structure can arise at some depth

below the surface even when the soil above may be fairly unaffected. The result is called a pan or a hard pan, a compacted layer at twenty or more centimetres below the surface that restricts downwards water movement even if the water has drained fairly freely initially. Pan formation can arise naturally in regions of very high rainfall when soluble mineral matter is precipitated from solution or when small insoluble mineral particles gradually accumulate. But pan formation can also arise as a result of soil being dug repeatedly and frequently to the same depth or, more especially, when a mechanical cultivator with L-shaped or similar tines has been used repeatedly. The result is a smearing of the soil and gradual development of the impervious layer.

As the soil structure deteriorates in any of the ways that I have described, so the subterranean environment becomes less favourable to earthworms and other forms of soil animal and ultimately there is established a rather sharp decline in the performance of the plants growing there. However, all of these adverse physical characteristics of the soil can and should be avoided, minimised or corrected and this is achieved by the use of appropriate methods of cultivation – although, as I have indicated, the cultivation itself can sometimes cause problems if used in the wrong way or at the wrong time. The term cultivation is a useful one to describe all of the physical operations on the soil that are the stock in trade of gardening – in particular, digging, forking, raking and hoeing in their various forms. For maintenance of good soil structure, digging is the most important of these operations and its tool, the spade, the most valuable.

The most important reason for digging is to break through the compacted layer at the soil surface, to break up the compacted soil mass that lies below the surface (and, if appropriate, to break through any impervious pan) and so enable air and water to penetrate freely. In doing this, the hard lumps of soil or clods are exposed to the actions of the weather. Secondary reasons are to facilitate the incorporation of organic matter at depth in the soil and to bury and thus kill annual weeds. Digging can be considered in two ways – in relation to the season and in relation to the depth to which it is performed. Because soil can only be dug by spade when fallow (free from plants) and not frozen, there is thus a choice between autumn and early spring. Both times are important. In autumn, soil should be rough dug; that is, the spade inserted approximately to its depth (usually called a spit) and the soil turned over but then scarcely broken down. During this autumn digging, manure or compost should be forked over the surface and roughly mixed with the soil. The water within the clods will be frozen during the winter and the soil will thus break down naturally into smaller pieces as a result of the increase in volume that occurs when water turns to ice. Even after this breakdown, however, the soil will still be in too coarse a condition for sowing seeds or planting plants when spring arrives. A second digging, either with a spade or with a fork, should be performed then, making liberal use of the back of the tool to break down the clods into finer pieces. If the soil is broken down in this way in the autumn, effort is being expended

One of the most important reasons for digging is to break up the compacted surface crust that is formed by the repeated beating action of rain. Such a crust impairs the penetration of air, water and nutrients.

wastefully (the frost will do the task for nothing) but, more importantly, the heavy winter rains beating onto the fine surface will bring about capping once again.

Although, so far, I have described digging in terms of the use of a spade, many gardeners have been attracted by the various powered rotary cultivators now available. I make some general comments about these machines on p. 174 but they have one major drawback as digging implements in encouraging the formation of an impervious pan, especially on heavy, clayey soils. Whilst a powered cultivator may provide the only means of working a large vegetable garden, it is important in these circumstances to do as the farmer does periodically when ploughing – to cultivate more deeply than usual by so-called sub-soiling every two or three years. The powered cultivator can also be a curse on land heavily populated with creeping perennial weeds such as couch grass for it will shred and disseminate their rhizomes. Despite the claims sometimes made that repeated rotary cultivation will eradicate couch grass, I have never seen any evidence for this.

Digging, as I have described it, is more correctly called single digging. The cultivation or disturbance of the soil is limited to the depth of the spade and the penetration of organic matter below this level is solely by the activities of earthworms and other soil animals. However, many plants, even many annual vegetables, root below this depth and from time to time it is important to work these lower layers too. The process is called double digging and it is best explained diagrammatically (see right). Double digging is hard work and it is unnecessary every season; one autumn digging in three should be adequate. There is one gardening practice that is dependent on double digging, however, and although almost as ancient as cultivation itself, it has become very popular in recent years. This is the deep bed technique which I shall return to shortly. First, however, I shall consider other basic cultivation methods.

The fork has two main uses in soil management,

Single digging

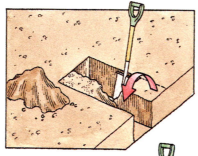
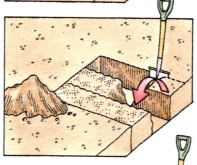
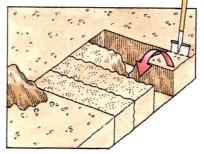

The cultivation is to one spade's depth only, the soil first removed from the trench being placed at the other end of the plot to refill the last trench.

Double digging

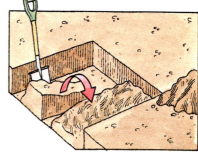

The basic system is the same as with single digging but the cultivation is to two spade's depth; the compacted soil is broken up with a fork and organic matter incorporated as the trench is refilled.

both of which I have referred to already. One is to facilitate a greater breakdown of the soil than is possible with a spade, the other is to lift and spread composts and manures. But even after the fork has been used in spring, the soil surface is still too coarse for seed germination and seedling growth to occur satisfactorily. It is said to have poor tilth. Partly this is because of the physical impedance that large pieces of soil and large air voids offer to the emerging seedling, but it is also because of the imperfect and uneven transmission of warmth that takes place through so heterogeneous a medium. Shortly before seed sowing or transplant planting, therefore, a rake should be used to remove the larger soil lumps still remaining and improve the tilth. On most soils, a spring-tined lawn rake performs this task admirably.

There are two common types of garden hoe – the draw hoe, which finds its main use in earthing up around potatoes and other crops to protect the crowns from frost or to encourage the development of blanched tissues, and the Dutch hoe. The latter is primarily a weeding tool, being used to sever the stems of annual weeds especially (see p. 171), but it is commonly used during the summer even on weed-free soil to loosen the surface and break up slight capping. In dry weather, this exposure of darker, moister soil undoubtedly enhances the appearance of beds and borders. Two contrasting beliefs have grown up around the practice of hoeing. One is that the exposure of moist soil enhances water loss, the second is that maintaining a loose, friable soil surface actually retards it. The first is true but only to the limited extent of water loss from the small amount of moist soil actually brought to the surface. Normally it is of no consequence, but in drought conditions when as much moisture as possible must be conserved, hoeing is generally to be avoided. Conversely, a dry soil surface layer will certainly act as a mulch and lessen water loss from below even if it is not hoed. (Mulching will be discussed further in this chapter). Moreover, whilst hoeing is an invaluable weed control technique, it is always an operation to be

undertaken carefully for damage can easily be done to shallowly rooted plants and, in the vegetable garden, onions in particular should never be hoed.

Apart from the periodic use of deep digging to break down any tendency for pan formation, there are two other occasions when it is important. The first is for trenching, a method of preparation used traditionally for certain types of plant that are usually said, in the evocative tones so beloved of some gardeners, to need a 'cool, moist root run'. Runner beans are the best example of such plants in the vegetable garden, and sweet peas in the flower garden. The principle behind trenching is to incorporate organic matter to a depth of about 30–50 centimetres in order to maintain soil moisture at a high level and encourage good top growth. The theory is sound, but care must be taken not to follow the advice so often given of layering compost or manure in the base of the trench. This is likely to encourage the development of a moist sump which may in turn lead to the establishment of the root-rotting organisms to which the pea and bean family is so prone. Organic matter must be incorporated uniformly throughout the depth of the trench and, on a light free-draining soil especially, care must be taken to grade the organically amended soil into the unamended soil at the sides of the trench, too.

Trenches are also used in the vegetable garden for the growing of blanched celery, although here the principle is different. The intention is partly to facilitate the blanching process by gradually backfilling the trench with soil as the plants grow, so denying light to their stems. (Despite the greater washing needed when using just soil, it is in practice better not to use cardboard or other sleeves around the plants, for these tend to encourage the establishment of slugs within.) The second reason for trenching celery is to enable the plants to be kept permanently moist; the celery naturally is a plant of wet, boggy places.

However, it is with the deep bed system of growing vegetables that deep, double digging is of especial importance. The principle of this approach

A 'cool, moist root run' is particularly favourable to sweet peas, and this can be obtained by using the technique of trenching when preparing the soil for cultivation.

exemplifies much that is sound in soil management for it seeks to minimise the compaction of the soil by avoiding as far as is possible walking or wheeling equipment over it. If this is done, the reason for routine annual digging is removed. But whilst it is still possible to add organic matter to the soil by mulching on the surface, the facility for incorporating organic matter at depth is removed too. Thus whilst the deep bed system offers the attractive option of not having to dig annually, it does replace this with the need to dig very thoroughly, deeply and doubly at the establishment of the bed and every six or seven years thereafter. The necessity of walking on the cultivated area is avoided by constructing it of such width that all parts can easily be reached from pathways at either side; in practice, a width of 1.0–1.2 metres is found to be most suitable. Thus weeds can be controlled by hoeing, and fertilisers and pesticides can be applied easily. If this distance still proves a stretch for sowing and planting, a temporary bridge comprising a plank supported on bricks, can be erected across the bed. On heavy, clayey soil, there is much to be said for hilling or raising the soil in the centre of the bed (so forming what is sometimes called a raised or Chinese bed) to improve drainage; this is a useful general cultivation practice in wet areas with heavy soils.

The relatively small amount of routine, annual cultivation needed with the deep bed system leads me on to an operation which always sounds so very appealing to many members of the gardening fraternity – minimal cultivation or zero tillage. Is this a euphemism you may wonder for doing nothing at all, or is it, like *Dianthus neglectus*, a name that merely tantalises to tempt the idle? It is in fact, like so much in horticulture, one name for several rather different techniques that had their origins in commercial practice and are not necessarily applicable to much of gardening. Perhaps the commonest form of minimal cultivation entails total dependence on herbicides for weed control (not a procedure that I would advocate in gardens) and the direct drilling of fodder root vegetables into the stubble left after cereal harvesting (not an operation with obvious gardening parallels). But there is one very sound gardening practice that certainly obviates the need for hoeing or any other method of annual weed control, which aids soil moisture retention so cutting down the amount of supplementary watering needed and which can minimise the amount of digging needed too if, as in the deep bed vegetable system, the soil is not walked on. The technique is mulching.

A mulch is simply a layer of material placed over the surface of the soil. In its traditional form, it is some form of organic material – a compost or manure of one of the types that I describe on p. 78. But in modern horticulture and gardening, it could also be of plastic sheet, either of familiar polythene style or of a type degraded in time by micro-organisms or ultra-violet radiation. Later in the book, I shall be advocating the use of mulch as a weed control measure and of course it is important to use substances that are themselves free from weeds or weed seeds. Correctly made compost and well-rotted animal manures are usually satisfactory in this respect. So, of course, are plastics, but whilst these will also aid moisture retention, they are of no value in improving soil structure in the way that an organic material does when earthworms and other animals drag it downwards. An organic mulch should preferably be at least 5 centimetres thick although it is interesting to note that in colder areas of the world, gardeners routinely apply mulches very much thicker than this to protect the soil from very severe freezing and provide insulation and frost protection for the crowns of perennials. In Britain, mulching of this thickness is scarcely necessary as an overall treatment although, of course, marginally hardy herbaceous perennials and shrubs can and should be protected with more localised application of organic matter mounded around them.

The overall mulch (or blanket mulch as it is sometimes called) should always be applied when the soil is moist, for its protective properties are as effective in keeping a dry soil dry as they are in keeping a moist one wet. The choice between an organic and a plastic mulch is likely to be made on the basis of the three criteria of availability, cost and appearance. Few gardeners are likely to be able to afford or obtain sufficient organic matter to enable them to use this routinely as a mulch in the vegetable plot as well as in more permanent plantings. Conversely, a plastic mulch is scarcely an aesthetically endearing embellishment to shrubberies and herbaceous borders although it is certainly useful in the vegetable plot. Whilst there is little if any difference in either the soil-warming properties (see p. 125) or the moisture retentiveness of clear, white or black plastic sheeting, it should be remembered that weed seedlings can grow and even flower and seed under a translucent clear plastic sheet, something impossible under a black one.

Whether used for mulching or for digging in and more thorough incorporation, it will be evident from what I have said that organic matter is an essential component of any gardener's soil management regime. There are some differences between the various manures and composts in respect of their nutrient contents (p. 79), but almost none in respect of their usefulness as soil amendment or, as they are sometimes known, conditioners. Any individual gardener's choice will, once again therefore, be dictated by cost and availability. Nonetheless, compost is within every gardener's reach and a correctly built compost bin should be a feature of every garden. There is nothing else in gardening that so beautifully exemplifies the harnessing of natural processes to the gardener's ends, yet there are few gardening operations that so many people seem to manage to do wrongly. In order to make compost correctly, it helps to understand the principles on which it is based and to understand why a compost bin is needed at all. Why cannot organic waste be used directly?

If fresh organic matter such as living plant material is placed directly on or in the soil, it will begin to decompose as it comes into contact with soil-inhabiting fungi and bacteria. Indeed, a definition of decomposition is the breaking down of organic matter into its constituent elements by bacterial or fungal action and it is important to appreciate how the micro-organisms achieve this. The organic matter provides them with a nutrient source; they feed upon it, at least in part secreting enzymes externally and so converting complex chemical constituents into simpler ones, some of which are used by the micro-organisms themselves in order to grow but others of which remain in the soil.

`   In fresh plant remains which are rich in cellulose, the ratio of the two important elements of carbon and nitrogen is about 33:1, whereas in the part-decomposed organic matter of compost or contained in soils as humus it is about 10:1. As fungal and bacterial activity increases on the plant remains, some of the carbon is utilised by them and some is lost as carbon dioxide but there is insufficient nitrogen to meet the demands of this vigorous microbial activity and more must therefore be obtained from elsewhere. A convenient source is the soil itself and although in time, as decomposition proceeds, the carbon to nitrogen ratio in the rotting vegetation falls towards 10:1 and the nitro-

gen becomes slowly mineralised, there is a temporary period when the soil nitrogen is depleted or unavailable for plants. Hence the reason for not applying fresh manure or unrotted plant remains directly to the soil in spring just before sowing and planting begins. And hence the reason for the compost bin in which the early stages of decomposition are facilitated by providing an additional source of nitrogen which is thus not removed from the soil. Composting can be conveniently defined, therefore, as the speeding up of the degradation of cellulose.

There are many ways of making compost and many different types of container in which to make it, but all depend on supplying a nitrogen source and carefully controlling the supply of water and air (and to some extent regulating the pH) so as to encourage fermentation and minimise putrefaction. The simplest technique and the one that provides the basis for most garden composting today depends on providing a good supply of air. I have set out below the practical operation of a modern compost bin system.

The two most common type of garden compost container. Each part of the double wooden bin should be about 1.2 metres cube, with slatted sides and an easily removable front. The plastic, barrel-style container is really only of value in very small gardens.

The most convenient size for a garden compost bin is a cube of about 1.2 metres, with slatted sides for aeration and with the facility for removing one side for ease of access. It has been calculated that adequate nitrogen will be supplied to the compost if the volume ratio of vegetable matter to fresh farmyard manure is about 3:1. But relatively few gardeners today are able to obtain so regular a supply of farmyard manure, so most find a scattering of nitrogenous fertiliser such as ammonium sulphate or a proprietary branded 'compost accelerator' a more viable proposition. Most types of organic debris can be added to the compost, although chicken carcasses and similar animal waste should not be used as they will attract vermin. Prunings and other woody material and tough vegetable stems should be shredded first – this will greatly increase their surface area and facilitate access by micro-organisms to the cellulose although the lignin component of the wood will be degraded very slowly. I have always found the inclusion of such shredded, tougher material to be extremely valuable in keeping open the structure of the compost and permitting the free gaseous exchange, so important in the early stages. Compost made with a high proportion (certainly more than about 30 per cent by volume) of soft green matter such as grass clippings or soft windfall apples will tend to encourage anaerobic bacterial decomposition (and attendant foul smells) from the beginning, instead of passing first through a largely fungal-mediated aerobic breakdown.

The original Indore system of compost making devised in the early years of this century in India by Sir Albert Howard required that the compost be turned at certain specified intervals to ensure that all parts experienced the correct sequence of organisms and temperatures (see below) but this presupposed that the whole container was filled at one time. However, most garden compost bins are filled piecemeal, and the simplest procedure therefore is to turn each batch of fresh material about three weeks after it is added and then to turn the entire mass as effectively as is feasible about three months

after the bin is filled. This will probably mean that the contents of the top quarter or so of the container are inadequately composted, but this material can be transferred to a second bin or used to start the first one afresh. (Two containers make planning compost manufacture a much simpler operation.)

Because of the piecemeal filling of a garden compost container, it is almost impossible to make compost quite as quickly or efficiently as in the three months that Howard originally demonstrated and most gardeners would be content with good results after about six months. This difference is largely because of the disruption that the periodic fresh additions make to the otherwise uniform natural succession of micro-organism populations and to the sequence of temperature changes that occurs within the compost as the micro-organisms perform their tasks. Because the chemical reactions taking place are exothermic (giving out heat), the temperature rises are considerable and even normal garden compost (at least in the centre) should reach 75 °C. (On p. 148 I shall discuss the implications of this for the composting of garden weeds.)

The value of a double compost bin cannot be overstated. It permits compost to be taken from one side while the other is still being filled with fresh material. In contrast, removing matured compost from the base of a single bin is almost impossible.

In mentioning the organic materials that could be added to compost, I specifically excluded leaves. Leaves are decomposed considerably more slowly than most other types of soft organic matter and they are broken down more by fungal than bacterial action. Their comparative slowness in decomposing is largely a reflection of their relatively large proportion of lignified conducting tissue in relation to their overall volume and also to their impervious protective cuticle. Whilst they can be added to compost therefore, they tend to obstruct the free flow of moisture and air. However, in the form of leaf mould, they do provide a very valuable mulching material and this is best made quite separately from compost in a leaf mould cage. Different types of leaf decompose at differing rates, and whilst it is impossible to quantify the time taken for each to reach the same state of decomposition, in general it may be said that oak leaves break down most quickly, followed by beech, ash and various maples. Large leaves with thick veins, such as chestnut and horse-chestnut, tend to be very slow at decomposing as are almost all conifer needles and other evergreen leaves.

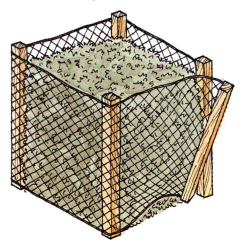

Leaves break down more slowly than other organic waste and are best stacked separately in an easily made leaf-mould cage, ideally about 1.2 metres cube.

Once the leaves have fallen from ornamental trees such as this Prunus 'Okame', they are no more attractive than other garden debris, though excellent as a mulch.

One organic material that has been a fairly recent arrival in most gardens is shredded, chipped or pulverised bark. It is produced as a by-product of the softwood timber industry and approximately half a million tonnes are produced annually in Britain alone. In its various forms, bark is a valuable mulching material and, although it has virtually no nutrient content, it has the merit of a very attractive appearance. But largely because of the cost of transporting and packaging such a bulky substance of fairly low density, branded bark is, and is always likely to be, fairly expensive. Understandably, gardeners with access to raw bark have been tempted to use it in their own gardens. This must be done with caution, however, for two main reasons.

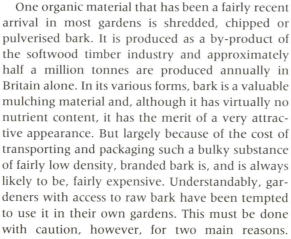

Chipped or shredded bark constitutes an excellent if rather costly mulching material best used for the more conspicuous, ornamental parts of the garden.

Some barks, especially those of spruces and Douglas fir, contain chemicals toxic to plant growth, especially various compounds collectively called terpenes. (Pine and larch barks generally have little such toxic effect.) By stacking the bark for several weeks, the content of the most important types of terpene is reduced by about 75 per cent, for the temperature very quickly reaches over 50 °C and the stack is invaded by thermophilic (heat-loving) bacteria. The stacking also results in a raising of the pH from about 4.5 to a more acceptable 5.5 and the elevated temperature has the incidental effect of eradicating the second potential problem – honey fungus (*Armillaria mellea*) which may be present in untreated bark.

After garden compost and farmyard or stable manure, peat is probably the organic soil amendment used most widely. It is less expensive than bark and makes an excellent soil conditioner. There are several different types of peat, some considerably cheaper than others, although much of the difference in cost arises from variations in the size and method of packaging rather than anything more fundamental. All peat is the part-decomposed remains of plant material; only part-decomposed because the plants lived and, more pertinently died, in waterlogged conditions conducive to limited anaerobic but little aerobic decomposition. There are two main types of peat available to gardeners, moss peat and sedge peat. In Britain, moss peat invariably means peat derived from species of the bog moss, *Sphagnum*, although in some parts of the world peat derived from other types of moss is important too. *Sphagnum* peat is fibrous and very spongy, generally brown in colour, with a very high water-holding capacity (up to ten times its own weight of water) and a pH ranging from about 3.5 to 4.00. Sedge peat is derived from sedges (species of *Carex*), heather (*Calluna vulgaris*) and certain bog grasses, especially common reed (*Phragmites australis*) and harestail cotton grass (*Eriophorum vaginatum*). It is more highly decomposed than *Sphagnum* peat, with a lower water-holding capacity (about three times its own weight), a dark

brown or black appearance and a higher pH. The pH range of sedge peat usually lies on the acidic side of neutral but in certain circumstances, as in the East Anglian fens, where the fen water has drained from calcareous rocks, it can actually be alkaline in reaction. In general, moss peat is to be preferred for the mixing of potting composts, but for large-scale addition to the soil for the improvement of structure, for mulching (but see p. 78) or for constructing a peat bed, sedge peat will almost always be the preference in most areas because of its lower cost and availability in bulk delivery.

So far, I have concentrated largely on describing organic matter in terms of the various types that can be added as mulch or that can be dug in. But there is another operation, very different in nature, that should also be considered. This is green manuring – actually growing an organic amendment *in situ*, usually in the vegetable garden. The principle of green manuring is to grow a crop and then instead of harvesting or removing it in any way, dig it straight into the soil, thus adding the nutrients that it contains and also supplying organic bulk to the soil. The crops most commonly used in gardens are rape, vetches, lupins, mustard, clover or comfrey (*Symphytum* spp.) and they are dug in, or 'turned under' when they reach a height of about 20–30 centimetres. The timing of sowing a green manure crop is critical if valuable garden space is not to be removed for a whole season. It is also very important that the crop chosen should be fast growing and, in most vegetable gardens, sowing the green manure after lifting early potatoes is the most straightforward plan. But is green manuring worthwhile and does it confer special benefits over and above more conventional ways of adding organic matter?

The actual organic bulk that can be added by green manuring is very small – it has been calculated that, on average, a green manure will contribute about 0.25 kilograms of dry organic matter per square metre, of which only a small percentage will actually form humus. Thus, the benefits must be looked for in terms of nutrient content and it has

Peat is widely used in potting composts, but unless roots are teased from the root ball when planting out, they may be reluctant to grow away from it and into the surrounding soil.

Leguminous plants are often used as green manure crops, making use, at least in theory, of the nitrogen accumulated in the bacterial nodules on their roots.

been argued that a green manure traps nitrogen that would otherwise be washed (or leached) from the soil by rain during the fallowing between normal crops. By using a leguminous manure crop, such as clover or lupins, there should also be some benefit from the nitrogen fixed by root nodule bacteria. But bearing in mind my comments on p. 55, it will be evident that the green manure will itself remove nitrogen from the soil as it decomposes, and unless the succeeding crop is sown very quickly (not always feasible or desirable), even the nitrogen within the manure crop may be leached away before it can be used. Moreover, a green manure crop will remove water from the soil, and in dry seasons a crop sown soon afterwards may actually experience difficulty in becoming established. There seems little if any evidence for the claim that deep-rooted manure crops such as comfrey actually tap otherwise unavailable nutrient reserves in the sub-soil. On balance therefore, the benefits to be derived from green manuring are arguable and the technique really has its main benefit in very exposed, windy areas with dry soils where the maintenance of vegetation cover for as much of the year as possible will minimise soil erosion.

Recently, gardeners may have encountered so-called biological or bacterial compost activators, with associated claims to the effect that they speed up and generally enhance the composting process. The principle appears to be one of adding a dried bacterial culture to the compost, but this seems of arguable benefit. Shortage of bacteria is seldom if ever a reason for unsatisfactory composting for there are always ample populations of bacteria on the plant debris that is added to the container. There was at one time a belief that it was necessary to add soil to compost, also in order to supply bacteria, but this too has been shown to be unnecessary. I believe therefore that perfectly satisfactory compost can be made without expensive bacterial additions, provided sufficient nitrogen is supplied for those bacteria already present for them to multiply.

Before leaving the subject of soil amendments, improvers or conditioners, I must draw attention to a type of substance that, although largely still at the experimental stage, may be expected to come increasingly to gardeners' attention in the future. To improve soil structure, it is necessary to enhance the processes of crumb formation. When organic materials are added to the soil, this process occurs essentially in three ways: through the temporary binding effect of soil micro-organisms, especially fungal mycelium; through the cementing action of microbially produced gums and polysaccharides (a type of carbohydrate); and through the cementing action of more resistant, stable humus components, aided by the similar action of some inorganic compounds such as iron oxides. The last mentioned provide most of the long-term stability of crumb structure. By using pure chemical compounds of similar glueing action, it should be possible to stimulate crumb formation on poorly structured soils without adding vast quantities of bulky material. Among materials that have proved themselves promising in this regard is a substance called cellulose xanthate, but it has not yet reached the garden market.

Almost everything that I have said about soil management, soil structure and soil amendments has a bearing on the subject of soil water. And whilst there can be no gardener (and certainly not one who has read the book this far) who will doubt the importance of water to plants, the concept of managing garden watering is still an alien one to most. In Chapter Nine I describe some of the watering devices now available to gardeners but here I shall concentrate on how and when to water most efficiently.

Water that falls onto the soil by precipitation is lost from it in three main ways. First by drainage into the sub-soil and then, via the bedrock, the underground water reservoir and natural watercourses into the sea. Second by direct evaporation from the soil surface. And third by evaporation at the surface of leaves after it has been taken up by the roots and passed upwards through the conducting tissues of the plants. The first two are purely physical processes, only affecting plants indirectly, but the third is, quite literally, vital. There must at all times be sufficient moisture available in the soil for plants to maintain this through-flow of water and its dissolved nutrients and, of course, also to maintain their structural turgidity. Ultimately, it is towards satisfying these needs that all water management is directed.

Improvement of soil structure as I have described will help matters in several ways. By simultaneously improving both porosity and crumb formation, it can either increase the moisture retentiveness of a sandy soil or decrease that of a clay, but in both instances it provides conditions more conducive to root growth and so will result in the plants being better able to utilise the available water. The improvements can be remarkable. In one series of experiments, for instance, garden compost was applied at the rate of 6 kilograms per square metre and dug in to a depth of 35 centimetres every autumn for ten years. It increased the amount of water available to plants from the soil by about 50 per cent. Thus incorporating an organic soil amendment will limit the amount of water that drains away into the sub-soil. Conversely, applying a mulch will increase moisture retentiveness by limiting evaporation from the surface. It has been argued that allowing the soil surface to dry and, in effect, to self-mulch is as effective as applying an organic blanket (see p. 59), but whilst a dry surface mulch certainly does limit water loss from several centimetres depth in the soil, it can still result in plants with very shallow roots, such as many annuals, suffering severe water shortage. Although the surface of a mulch itself can dry out, its physical nature means that it will do so much less readily than the surface of the soil. And, of course, as plants do not root in the mulch no damage to them will ensue from any surface drying. (In passing, I should add that of all the organic materials suitable for mulching, peat is in many ways the least satisfactory, for once its surface dries it is notoriously hard to re-wet and a dry peat

mulch, far from acting as a sponge, can actually throw off water applied to or falling onto it.)

There are several features of plant husbandry that can either influence directly the amount of moisture in the soil or, more commonly, enhance the ability of the plants to make use of it. I have already mentioned several times the loss of water through the leaves, and this is very much greater than is generally appreciated. In hot summer weather, for instance, it can be at least 5.5 litres per square metre of leaf area per day. Clearly, this could be limited by growing fewer plants and I offer this as something for serious consideration. I referred to the detrimental effect that green manure crops can have on soil moisture, but weeds offer another very common and quite avoidable route for the reserves to be depleted. Among all the other reasons for keeping weeds under control therefore (see p. 140), the conservation of water is a very real one, especially in dry weather and on light soils. But it should be practicable actually to consider your numbers of crop plants (especially of vegetable crops) too. A mature cabbage plant probably offers about 1 square metre of leaf area. For every surplus cabbage that you grow therefore, an unnecessary volume of water that could represent as much as 5.5 litres per day is being lost and denied to your other plants.

Plant spacing has water management implications too. The closer plants are placed together, the less will be the volume of soil that each root system has available and hence the smaller will be the volume of soil moisture that each plant is able to tap. Ultimately, this will have its results in the plants actually attaining a smaller size and although there may be a few occasions when this is desirable (pp. 107–8), in general, most gardeners actually prefer their plants to achieve as much as possible of their growth potential. I have seen the argument made that close spacing means a greater soil cover and thus less direct evaporation from the soil surface but this is specious reasoning, for the loss from leaf surfaces is very much higher than that from the soil.

Ultimately, no soil moisture conservation

Much careful study has gone into the science of plant spacing, and, for each type of crop, optimum distances between plants have been calculated. Such spacings are a compromise between those giving theoretical maximum yields from a given area and those giving the most practically useful plant size.

measure will be fully adequate to satisfy the water needs of your plants in the summer, and artificial watering must supplement this. But conservation and careful use of supplementary water is important too. Partly this is to ensure that national water resources are not wasted but it is also because, despite what is often believed, most plants do not actually grow better, the more water they are given. Commercially, of course, the volumes of water used for crops and the financial implications of this usage are enormous, and it is not surprising that a great deal of research has been performed to establish the optimum water requirements and watering regimes for all major fruit and vegetable crops. I cannot do justice here to all of these findings and, in any event, it is unrealistic to expect gardeners to so precisely regulate watering from one crop to another, but if you appreciate certain general principles, it should enable you to plan your water usage fairly meaningfully.

Whilst all plant tissues contain protoplasm which is composed largely of water, the actual response

displayed when supplementary water is applied is not uniform – some tissues react more than others. In general, the green tissues (leaves and stems), display the quickest and greatest effect and there are many occasions when this has a direct and obvious benefit – lettuces will heart up and spinach produce more edible foliage, for instance. But where it is fruits, roots, tubers or flowers that are the objects of cultivation, there may not be a direct benefit as a consequence of the enhanced leaf growth. In some plants, leaf growth may continue apace at the expense of other more desirable parts and plants do differ rather widely in this respect. Assuming however that you do not have a limitless water supply and limitless time to apply it, and assuming that you will not lose too much sleep if your plants do not produce the maximum theoretical response to additional watering, I would commend the following guideline: give supplementary water at the time of the season when the parts that are the objects of cultivation are beginning to mature. Thus, leafy vegetables such as cabbages or lettuces should be watered as the heads fill, carrots and other root crops as the roots begin to swell, potatoes as the tubers start to form (approximately at flowering time), soft fruit as the fruits set, and annual flowers as the buds expand. In most seasons, most perennial food crops (with the exception of soft fruit) and most perennial ornamental plants will perform satisfactorily even if they cannot be given any additional water, provided the soil has been well tended and mulched.

But how much additional water should be given, bearing in mind my earlier comments that some plants respond adversely to overwatering? There are several ways of considering the quantity of water – in relation to the area of soil over which it is applied, in relation to the depth of the soil that is wetted to obtain a given moisture content or in relation to the time and flow rate of the delivery system. Think first of supplementary watering in comparison to rainfall: 2 inches of rain represents 1 gallon per square foot (or, in less easy to remember metric equivalents, 5 centimetres represents 0.6

litres per square metre). In relatively low rainfall areas, such as the east of England, 5 centimetres also represents the approximate average monthly rainfall (in these low rainfall regions the month to month variation is generally very much less than in high rainfall areas where the wettest month may have three times as much rain as the driest). Anyone living in such a dry area, especially on a light soil, realises that in an average year the high evaporation rates in summer mean that many garden plants find the natural water supply insufficient; and in a dry season, rainfall is a woefully inadequate supplier of water. There may be gardeners who will take their own rainfall measurements or obtain them from a local meteorological station and calculate the amount of water, week by week, that is needed to make up any natural deficit and supplement it by an appropriate amount. Then, by using a modern water-flow meter and sprinkler system (p. 165), they water their garden uniformly. I don't doubt that these gardeners are in the minority, and I know also that in dry areas such ad lib use of sprinklers will almost certainly be legally prohibited.

I suggest, therefore, two alternative courses of action. The first applies where use of a sprinkler is permitted and water is freely available. Assume that, on average, crop plants have what I call an active productive season (the period during which most of the flowering, fruiting or other crop development is taking place) of about four months. If you are able or wish to water them once only (assuming that they are already established and past the seedling or transplanting stage), apply water at the rate of about 25 litres per square metre at the period most critical for most of the plants (see p. 41). Alternatively, apply about 15 litres per square metre once a month or 10 litres per square metre once a fortnight. This will not ensure that the soil is constantly wet down to rooting depth, but it should ensure that most of the water is present where and when it is most needed. Of course, a large proportion of the water will fall onto bare soil and will evaporate from it, and when and where

water supply is officially regulated, this system is unacceptably wasteful (and, if the water must be supplied by watering can, impossible too). Then you should restrict your watering to vegetables, annual flowers and possibly to soft fruit, and limit the watering to the area of soil close to the plants — approximately the extent indicated by the leaf spread. Use the rates and frequencies that I have indicated however, but scale down the areas and volumes. This would mean that if cabbages, once established, were to be watered once only, this should be done as the heads are beginning to swell and that water should be applied at the rate of about 3 litres to an area of about 0.9 square metres (about one square yard) around each. If they are to be watered monthly, then each should be given about 1.5 litres at each watering and, if fortnightly, about 1 litre. In the latter instance, a plot of three dozen cabbages would be given a total of 36 litres or about five 2-gallon bucketfuls every two weeks, surely a manageable undertaking.

Always try to water plants in the evening so that the applied water has a chance to soak into the root zone with little loss to evaporation. It is sometimes suggested that plants should never be watered in full sun, and whilst this is certainly wasteful of water, only in very hot conditions and when water is actually poured over plants with tender leaves and open flowers is much actual damage likely to ensue. Much water is wasted every year on lawns and I am constantly amazed and saddened at the way gardeners spend time, effort and valuable water in trying to ensure that their lawn stays green through a hot summer when their vegetables, soft fruit or ornamental gardens suffer. Lawns, despite their relatively shallowly rooting, have quite astonishing powers of recovery and even after the severe droughts of the mid-1970s in Britain, I doubt if many actually died. Plants grown in containers should, of course, never be denied watering for they have no reserves at depth to tap. Tubs and window-boxes should be watered once a day and hanging baskets in sunny positions preferably twice.

CHAPTER FIVE

Foods and feeding

O f all the differences between plants and animals, the one that is most directly related to
the way that plants feed is their inability to move –
they cannot walk or run from place to place in
pursuit of a meal. Whether the ability to obtain all
their nutrient requirements while still remaining in
one place is a consequence of, or a reason for, a
stationary existence is for evolutionists to ponder.
Nonetheless, plants are certainly restricted, for
being literally rooted to the spot means that they
have only two sources of raw materials, the soil and
the air, and both are used to the full.

The most important chemical element for any
living organism is carbon and it is the possession of
compounds containing carbon that sets life apart
from inanimate objects. These carbon-containing
substances are generally called organic compounds
and include proteins, carbohydrates and fats. Plants
have a convenient source of carbon at hand in the
form of the gas carbon dioxide which accounts for
0.03 per cent of the atmosphere. Close at hand in
the atmosphere, too, is water vapour, and by means
of a simple chemical equation, water plus carbon
dioxide can equal carbohydrates, more complex
chemicals that are used as a means of storing
energy. This energy is required and used initially to
power the chemical reactions in which the carbohydrates, which include sugars, starch and cellulose, are made, but it can be released when the
carbohydrates are broken apart again into their
constituents. The manufacture of carbohydrate in a
plant is by the unique process of photosynthesis; its
subsequent breakdown is by the much more widespread reaction of respiration, essentially the same
process as takes place in animals and other organisms – although they, of course, must first obtain
their carbohydrates by eating plants. The energy

*The food requirements of
plants in the full flush of
summer growth are
prodigious and gardeners
far too often make the
mistake of believing that all
this growth can be sustained
by water alone.*

released during respiration is used to enable yet further chemical combinations to take place and so form proteins, fats and other substances required for manufacturing new cells and tissues which enable the plant to increase in size.

Photosynthesis, as its name suggests, is a chemical manufacturing process that takes place in the presence of light – it is sunlight that provides the energy to combine the water and carbon dioxide, the reaction taking place in those cells that contain chlorophyll-bearing chloroplasts. The role of the green pigment chlorophyll is usually said to be to trap the sun's energy and this occurs by the absorption of light and then a series of transfers of the energy from one type of chlorophyll molecule to others. Complex organic phosphate-containing compounds are important too.

Quite clearly, carbohydrate manufacture by photosynthesis and energy release by respiration are complex processes but they are also robust enough to function well in the numerous and extremely diverse environments that plants have come to occupy. They might not seem at first sight to be processes that gardeners can influence directly, but plants will not grow satisfactorily if they do not have the necessary raw materials – light, a supply of fresh air and the wherewithall to manufacture chlorophyll. Adequate temperature is important too, for quite apart from the purely physical damage that can be caused to tissues from freezing or even chilling, all chemical reactions, photosynthesis and respiration included, are slowed down by a decrease in temperature. Denied any of these, a plant will gradually decline.

However, the products of photosynthesis are only the first part of the story of plant nutrition and it is in the provision of the second that gardeners can and should play a much more direct role. The second source of raw materials, the soil, provides a plant with its mineral nutrients, and without these the range of chemicals that can be manufactured would be limited indeed – the proteins, fats and other substances that I have mentioned all contain a very much greater diversity of chemical elements than the hydrogen, oxygen and carbon that are obtained from the air.

There are well over a hundred chemical elements known to exist on earth, but plants require only a small number of them. The most important are nitrogen, phosphorus and potassium, generally called the major elements, followed by calcium, magnesium and sulphur (the minor elements). Iron, manganese, boron, copper, molybdenum, zinc, sodium and chlorine are only required in extremely small amounts and they are known accordingly as trace elements. Before describing the specific roles that each has to play, I can make some general comments about the way that the presence of different elements varies in the soil and how presence is not necessarily synonymous with availability to plants.

The relative amounts of different elements in the soil are primarily a consequence of the type of rock from which the soil is derived and the minerals that the rock contained. Limestone and chalk, for instance, contain the element calcium but very few others, whereas granite is rich in a wide range of minerals and many different elements. Whilst many soils contain adequate reserves of most if not all of the elements that plants require, growth may be less than satisfactory in them. The reason is that reactions take place between the chemical compounds in soil and in certain conditions, especially when the pH is high, the required elements may be converted to a form which renders them less soluble or otherwise incapable of being taken up by roots. I have indicated in the diagram on the right the way that pH affects the availability to plants of nutrient elements and it will be seen that, with the exceptions of calcium, magnesium and molybdenum, virtually all elements are less available in alkaline conditions. The plants that we call calcicoles (p. 42) have adapted to this state of affairs, but most garden plants are not calcicoles and will thus be deficient in many of their nutrients in an alkaline soil. The most significant element in this respect is iron, for iron is required by plants in the formation of chlorophyll (although it is not, as

The relative abilities of plants to take up certain mineral elements from the soil at different pH levels – the thickness of the bands indicates relative ease of uptake.

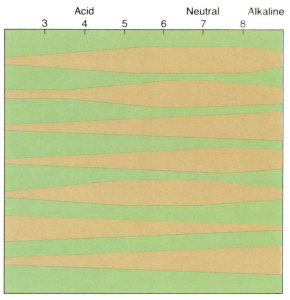

Acid Neutral Alkaline
3 4 5 6 7 8

Availability to plants of the elements nitrogen and sulphur

Availability to plants of the elements phosphorus and boron

Availability to plants of the elements calcium and magnesium

Availability to plants of the element potassium

Availability to plants of the elements copper and zinc

Availability to plants of the elements iron and manganese

Availability to plants of the element molybdenum

sometimes believed, a constituent of chlorophyll itself). Without chlorophyll, of course, a plant will not only turn yellow but will ultimately die; therefore some means of circumventing the problem is needed. The answer for gardeners is to supply iron in a chelated or sequestered form in which the element is bonded onto an organic molecule. This will not allow it to combine chemically with other elements in the soil in the way that inorganic iron does. The symptoms of most nutrient deficiencies (or unavailabilities) are fairly diagnostic and it is useful for gardeners to be able to recognise them. Accordingly, I have indicated in a table overleaf the specific effects that shortages of them induce and suggested the appropriate remedies.

You will already appreciate that plant nutrients are taken up from the soil in water solution. But gardeners often ask how the chemicals actually manage to pass from the soil into the cells of the root while, at the same time, the liquid contents of the root do not all fall out. The answer for water uptake lies in a physical phenomenon called osmosis, a word derived from the Greek meaning thrust or push. It is a simple law of physical chemistry that when there are two solutions of differing concentration, separ-

ated by a semi-permeable membrane (one that does not form a total barrier but has minute pores in it), molecules of water will pass through the membrane into the solution of higher concentration, and they will continue to do so until the concentrations of the two solutions are the same. The osmotic concentration of the solutions in the root is higher than that of the water in the soil, the root cytoplasm is a semi-permeable membrane, and so water molecules pass inwards. The uptake of minerals is more complicated and less understood although the first stage is by diffusion through the fully permeable root tip cell walls into the intercellular spaces.

One further important point about plant nutrients and nutrient uptake should be stressed. In order for plants to assimilate them and use them in the building of new chemical substances, the nutrients must be in chemically simple form. Take magnesium for example, a common enough element that is believed to comprise about 2.7 per cent of the earth's crust. Naturally, it occurs principally as magnesium carbonate in the minerals magnesite and dolomite, as magnesium sulphate in epsomite, kieserite and other more complex potassium and calcium-containing minerals, as magnesium chlo-

Symptoms and treatment of the commonest plant nutrient deficiencies

Nutrient	Plants commonly affected	Symptoms	Treatment
Nitrogen (N)	Most, but especially common on brassicas and other leafy vegetables	Leaves reduced in size, pale or reddish-purple coloured – first on older leaves. All growth weak, fruit and other produce reduced in size and number	Apply nitrogen-containing fertilisers routinely each season, especially in advance of planting
Phosphorus (P)	Most	Leaves small, may drop prematurely, sometimes dull blue-green or bronze – first on older leaves. All growth weak, flowering and fruiting delayed and reduced	Use balanced NPK fertilisers each season. For specific phosphorus deficiency, apply superphosphate or bone meal close to plants
Potassium (K)	Beans, flowering perennials, fruits, leafy vegetables, potatoes, tomatoes	Leaf tips and margins scorched – first on older leaves. Often general weak growth	Use balanced NPK fertilisers each season. For specific potassium deficiency, apply potassium sulphate in advance of sowing or planting
Calcium (Ca)	Apples, brassicas, carrots, celery, chicory, lettuces, peppers, potatoes, tomatoes	Leaf tips and growing points blackened; bitter pit spotting in apple flesh; blossom end rot on tomatoes; internal browning on Brussels sprouts; leggy shoots on potatoes; tip burn on lettuce	Apply lime in accordance with recommendations for soil type (see Fig. 14). Spray affected apples and tomato fruits with calcium chloride solution (2g/l)
Magnesium (Mg)	Apples, bedding plants, some brassicas, lettuces, potatoes, roses, tomatoes	Leaves with yellow interveinal marbling – first on older leaves and sometimes confined to central part of leaf	Use magnesian limestone; also magnesium-containing fertilisers on roses and other shrubs
Iron (Fe)	Azaleas, camellias, ceanothus, chaenomeles, fruit trees, hydrangeas, rhododendrons, roses, soft fruit (especially raspberries and strawberries)	Leaves yellowed uniformly but with dark green veins	Apply sequestered iron
Manganese (Mn)	Beans, beetroot, brassicas, fruit trees, parsnips, peas, potatoes, soft fruit, spinach	Leaves with interveinal yellowing, often with dead patches. Brown marsh spots on pea cotyledons	Avoid over-liming on susceptible sites. Spray affected plants with manganese sulphate (1.5g/l)
Boron (B)	Apples, beetroot, brassicas, celery, lettuces, sweet corn	Very similar to calcium deficiency; also corky patches in apple fruit and other plant structures; sweet corn leaves striped and growing point dies	Rake in borax at planting time at 3g/m^2
Molybdenum (Mb)	Cauliflowers	Suppression of leaf blade growth giving whip-tail effect	Use lime to raise pH on acid sites. Water seedlings with ammonium molybdate (5g/l/m^2)

ride in kainite and to a small extent in minerals containing magnesium oxide. None of these magnesium compounds can be used by plants directly. In the soil water, however, they are broken down chemically to varying extents to release magnesium ions.

It is now appropriate briefly to outline the relative importance and the use to plants of the individual mineral nutrients. Nitrogen is generally considered the most important plant nutrient for it is a major constituent of protein, of protoplasm and other plant components, and it is also incorporated into many other organic chemical compounds. In gardening terms, it is always associated with leafy growth rather than flower and fruit development; too much nitrogen can delay or even inhibit flowering. Its main source is dead organic matter (the element therefore recycles constantly through living organisms), but nitrogen is uniquely different from all other soil elements in that it also passes into mineral form in the soil from the air. The roots of some plants, especially those in the family Leguminosae (peas, beans and their allies), bear small nodules, and within these species of bacteria belonging to the genus *Rhizobium* live in an interdependent or symbiotic relationship with the plants. It seems that the process, called nitrogen fixation, involves in part the formation of ammonia and amino acids, but it may differ between organisms. Ultimately, nonetheless, the nitrogen finds its way into the soil in a form that can be used directly by plants. The special importance that leguminous plants thus have in cropping patterns is considered on p. 95. Nitrogen disappears rapidly from the soil by leaching and almost any soil can become nitrogen deficient after repeated cropping although not surprisingly the effects are almost always most serious on those with a low organic matter content.

Phosphorus is involved in many aspects of plant growth and is a constituent of many different carbohydrates, proteins and fats. It has a special importance in the ripening of fruits and in the ripening and germination of seeds, and it is also associated with the efficient functioning of nitrogen in the plant. Gardeners have long recognised the value of phosphorus in the encouragement of root growth. The main sources of phosphorus for soils are complex phosphate minerals such as apatite, a complex of calcium phosphate, fluoride and chloride, and the element is quickly leached in regions of high rainfall and from acid soils generally although it can also be deficient on heavy clays.

Potassium is an enigmatic mineral, well known to be of great importance for good growth and also in encouraging flower and fruit development. Yet it is of somewhat uncertain function although it is clearly involved in part in photosynthesis and also in the control of water loss from the leaves. Most soil potassium originates from the potassium and aluminium silicate called orthoclase feldspar, a major constituent of granite and other igneous rocks. It is most commonly deficient on light, sandy soils and also commonly on chalky and peat soils with a low clay content.

Scorched and browned leaf margins are a good indication of a deficiency in plants of potassium, an element especially important for flower and fruit development.

Blossom end rot on tomatoes is the most common symptom in gardens arising from a shortage of calcium in the plant's tissues. Such a shortage is often brought about or exacerbated by dryness at the roots.

The effects of magnesium shortage are sometimes confused with the symptoms of virus infection although on potato foliage, the interveinal yellowing and browning is diagnostic.

Calcium has special significance in the structure of cell walls, being an essential part of the pectate molecules that help form the central wall membrane, and it is important for the correct functioning of the growing points of shoots and leaves and also to some extent in root growth. It has a special role both in plant tissues and in the soil in relation to the availability of other nutrients. As I explain on p. 80, nitrogen in the form of ammonia can limit the availability to plants of calcium, but excesses of potassium and to some degree of magnesium can restrict it also. Calcium deficiency can also be exacerbated when plants are short of water, presumably because calcium is relatively immobile in the tissues and a free flow of water is necessary to transport it from old to young tissue. Within the plant, a deficiency of boron can also lead to calcium shortage because boron is required for some aspect of the movement of calcium. The main sources of calcium for the soil are limestone, chalk and other rocks containing calcium carbonate, but there is a close relationship between calcium content and pH in many soils; in Britain an alkaline soil is usually one rich in calcium, too. Many plants, called calcicoles, require the particular iron-calcium relationship that occurs in calcium-rich soils whereas others, known as calcifuges, require a different relationship between these elements and do not thrive in such conditions. By inference, therefore, calcium shortage may be expected in soils derived from acid peats and in those that originate from acidic rocks like granite that are naturally low in the element.

Magnesium is a constituent of chlorophyll and therefore has a very special part to play in plant nutrition although it has other functions relating to the plants' chemical processes or metabolism too. I indicated its main mineral sources on p. 66 and its

common occurrence in dolomitic limestone means that its availability is associated closely with that of calcium. High levels of soil potassium can render it deficient and, as it is readily leached, it is also often in short supply after very wet seasons. Sulphur is present in inorganic minerals such as gypsum (calcium sulphate) and iron pyrites (iron sulphide), but also in organic matter for it is an essential constituent of many amino acids and thus plays a vital role in protein structure.

I have already discussed the important role of iron in the plant and how its availability may be impaired in alkaline soils. Iron is rarely deficient in soils and it is the cause of the typical reddish brown colour that many soils display. It is in fact, after aluminium, the most abundant metallic element in the earth's crust and soil iron usually originates from iron oxides like magnetite and haematite and the carbonates like siderite.

Manganese is another element of vital but uncertain function although, like iron, it appears to be associated with the formation of chlorophyll. The main sources of soil manganese are various manganese oxides, but it is commonly in short supply, especially in organic soils, in wet alkaline soils and on poorly draining sites generally.

Boron is rarely deficient in the soil but it is commonly unavailable to plants as it occurs in minerals such as tourmaline which plants are unable to utilise. High alkalinity also renders it unavailable. The function of boron in the plant is associated closely with that of calcium and a deficiency results in damage to the growing points. Boron may also aid the translocation of sugars, but its precise role is the most imperfectly understood of all the most important nutrients.

Copper is of partly uncertain function but has considerable importance as a constituent of several

Lime-induced chlorosis is the term commonly given to the effects of iron deficiency. On the raspberry plant, below left, the leaf as a whole is yellowed but the veins remain characteristically dark green.

In this factorial experiment, lettuce plants have been grown with relatively increasing deficiencies of both calcium and boron, two elements whose effects are often closely interrelated in plants.

important enzymes (chemicals that assist in chemical reactions). It usually originates in such minerals as chalcopyrite in the form of sulphides, and on extremely acid or extremely alkaline sites it can be rendered unavailable. Molybdenum is important for its role in rendering nitrogen available to plants through the chemical reduction of nitrate salts to ammonium and also for the nitrogen fixing by root nodule bacteria. Its deficiency is very rare although in acid soils it can become unavailable through becoming chemically fixed with other soil elements.

Zinc in soils usually originates in minerals such as blende, smithsonite and hemimorphite (as sulphides, carbonates or silicates) and it may become unavailable to plants on light, sandy, alkaline sites. Zinc appears to be related to auxin metabolism in plants in several different ways – although it does not seem to be required for auxin synthesis, the presence of zinc is necessary for auxins to be maintained in an active state.

Sodium is present in such minerals as common salt (sodium chloride) and mirabilite (sodium sulphate), and in so-called sodic soils it can occur in very high concentrations. Its role in plants is uncertain although suggestions have included a function in photosynthesis, in carbon dioxide assimilation into tissues or for the activation of some enzyme systems. Many plants of maritime origin (such as beets) have a particularly marked requirement for sodium – not surprisingly since it is sodium chloride that gives sea water and coastal soils their saltiness. Chlorine is present in metal chlorides such as sodium chloride. Its deficiency has various effects in different plants and it is believed to be a co-enzyme, essential for the light-induced chemical reactions in photosynthesis.

In an ideal world and an ideal ecosystem, what goes out from the soil comes back again when plants die and so there is always enough nutrient for new plants to grow and flourish. But the ideal world and the ideal ecosystem is an arcadian impossibility and you will recall from p. 44 that in a real ecosystem organisms, raw material and energy move in and out. In a garden they move out considerably more swiftly than they move in as we harvest our cabbages and runner beans, sweep away the leaves of autumn and cart lawn mowings to a municipal tip. Nutrients in the soil become depleted and it is the role of fertilisers to replace them. Fertilisers are sources of plant nutrients that we, rather than nature, supply. They vary one from the other in the effect they have on plant growth in two main ways. The first is a consequence of the fact that they do not generally contain nutrients in a form that plants can utilise directly. The ease and rapidity with which the necessary chemical and to some extent physical changes take place before fertilisers can be used are very important therefore. The second way in which fertilisers vary is in respect of the differing proportions they contain of each type of major and minor nutrient and trace elements.

Rapidity of chemical breakdown and hence of availability to plants is partly a function of the ease with which micro-organisms in the soil can degrade fertilisers. But more generally it is an overall function of the solubility in water of the various nutrients, for remember that it is in a water solution that mineral nutrients actually pass into the roots. Varying water solubility among chemicals is a common feature of everyday life; salt dissolves easily in cold water, sugar in hot and sand not at all. I have indicated opposite the water solubility of some of the major chemical compounds occurring in fertilisers. By this token, therefore, sodium nitrate is a quick acting nutrient source and tricalcium phosphate a slow one. But sodium nitrate is only quick acting if plants are there at hand to take it up into their roots. Its high water solubility means not only that it will dissolve quickly in the soil water but also that it will very quickly be washed or leached from the soil as that water drains away. Hence a highly soluble or quick acting fertiliser is only of real value at a time of year when plants are able to make use of it before it disappears. The extreme form of a quick acting fertiliser is that generally called a liquid feed which is dissolved in

Water solubility of some of the major chemical compounds occurring in garden fertilisers

Compound	Solubility in water (grams per 100 millilitres)
Ammonium nitrate	118.3[a]
Ammonium sulphate	70.6[a]
Calcium carbonate	0.00153[b]
Calcium sulphate	0.209[c]
Tricalcium phosphate	0.002[a]
Potassium chloride	34.4[a]
Potassium sulphate	12.0[b]
Sodium nitrate	92.1[b]

[a] at 0 °C
[b] at 25 °C
[c] at 30 °C

water even before it is applied. Sprayed onto the leaves rather than the soil (to make use of their limited ability to take up liquid through the cuticle), a liquid feed provides the fastest possible means of delivering nutrient into a plant's tissues. Conversely, of course, a chemical such as tricalcium phosphate provides a steady source of nutrient over a long period of time and will persist even through periods of heavy rainfall. However, such slow acting nutrient sources provide relatively little nutrient at any particular moment; this is a disadvantage when plants are growing very rapidly and have a high nutrient demand in mid-summer.

Unfortunately, fertiliser technology is not quite as simple as I may have suggested. While it is perfectly possible to use pure chemical compounds of the types I have mentioned, each compound will generally supply only one or perhaps two of the nutrient elements that plants need. Ammonium sulphate, for instance, contains nitrogen, hydrogen, oxygen and sulphur and when dissolved in water these are liberated; however, only the nitrogen is actually used by plants. Although there may be rare occasions when plants have abnormal needs for nitrogen, potassium or another single element, it is a reflection of good fertiliser practice that such deficiencies should not arise because the feeding has been with a balanced mixture containing appropriate amounts of all required elements. But what is appropriate you may wonder and how can one fertiliser be correct for different types of plant on soils having different natural chemistries? In short, the answer is that it cannot, which is one reason why so many different types of fertiliser grace the shelves of garden centres. Most of them are called compound fertilisers, in contrast to the more or less pure chemical compounds which are popularly called straight fertilisers. Compound fertilisers are balanced blends of different chemical ingredients, selected for the needs of particular plants or gardening applications. When we speak of a fertiliser as balanced, this generally means that it has been formulated with special reference to the nitrogen needs of the plants for which it is intended, nitrogen being, as I have explained, not only the most important single nutrient element but also the one most rapidly lost from the soil. Thus, primary consideration has been given to the nitrogen content, and the amounts of phosphorus and potassium have been balanced accordingly.

The way in which the nutrient content of fertilisers must be described on package labels is strictly laid down by British law under the Fertilisers and Feeding Stuffs Regulations (comparable legislation exists in several other countries) and it is very useful to be able to understand what at first sight appears to be chemical mumbo-jumbo. First you will see a statement of the type of fertiliser – compound or simple, for instance. There then follows the percentage by weight of the three major plant nutrients, in the order nitrogen (N), phosphorus (P) and potassium (K) – 10:10:15, for example. More detailed analysis follows in which the nitrogen content is expressed simply as the percentage by weight of the element, but the percentages of phosphorus and potassium (at least in Britain) are expressed in terms of the weights of their oxides. Thus phosphorus is expressed as phosphorus pentoxide (P_2O_5 – commonly called phosphate), and potassium as potassium oxide (K_2O – commonly called potash) with a further value, in brackets, of the actual elemental amounts available

Vigorous, luxuriant growth often implies the use of fertilisers in the garden. The choice between 'organic' and artificial feeding is an individual one, but an understanding of the scientific debate can only help to clarify the gardener's perspective.

to plants. (To convert weights of phosphate and potash to weights of elemental phosphorus and potassium, multiply by 0.436 and 0.83 respectively.) An additional subdivision is used for phosphorus which expresses the proportion of phosphorus pentoxide soluble in water so that it is possible to judge the amount more or less immediately available for growth. (Even the water-insoluble proportion will become available eventually as it is dissolved slowly by soil acids.) The percentages by weight of any other important elements contained in the fertiliser, such as magnesium or boron, are also included on the label. One additional point must be made. It is the relative *proportions* of N, P and K in a fertiliser that are most important, not their absolute amounts. A fertiliser of composition 2:2:5, for instance, is relatively just as potassium rich as one of 8:8:20; you will simply require four times as much of it to give your plants the same amounts of nutrients.

I have summarised on pp. 80–2 the salient features and relative merits of the most important fertilisers readily available to gardeners and have also summarised on pp. 82–6 what I believe to be the most useful range for the average amateur gardener whose needs are inevitably less demanding than those of the professional. But I must first discuss several general features of fertilisers, beginning with perhaps the most contentious aspect relating to fertiliser choice; the relative merits of so-called organic (or natural) and so-called inorganic (or artificial) types. To a chemist, an organic substance is one containing the element carbon, but the gardening fraternity seems to prefer the older meaning of 'derived from some living or once-living organism'. Such a fertiliser therefore contrasts with an inorganic substance which originates in some other way, usually, although not necessarily, an artificial manufacturing process. As will be evident from what follows, most inorganic fertilisers are of rather precisely definable composition whereas most organic ones are not, although this has rather little practical significance in gardens. Many organic fertilisers do contain small amounts of minor and trace elements in addition to their major components, although I find it hard to credit this a significant benefit for the reasons I outlined on p. 72. Most inorganic fertilisers are highly soluble and thus fairly fast acting although the development of coated fertiliser granules or pellets in recent years has added a controlled solubility and controlled release capability to the range. And, whilst most organic fertilisers are more insoluble and slower acting or, as it is called, slower release, some can be rendered fairly fast acting if they are ground to a fine powder which increases the particle surface area and hence the solubility of the material.

Understandably and rightly, gardeners are aware of the need to minimise both contamination and destruction of the natural environment and it is on the usage of fertilisers in particular that much of conservationists' venom has been directed in recent years. There are two aspects to this. The first is the claimed contamination of the environment in general and of water supplies in particular by excessive usage of artificial nitrogenous fertilisers on the land. The use of artificial fertilisers overall in gardens accounts for only a minute amount of Britain's total national consumption (about 1.5 million tonnes of nitrogen) and the fertiliser manufacturing industry is insistent, moreover, that claims of cancers, especially gastric cancer, arising as a result of people having consumed abnormally high levels of nitrates are ill-founded. They point to the fact that British agricultural soils naturally contain about 150 times the amount of nitrogen that is added artificially in fertilisers each year, and that we all eat about 16 grams of nitrogen every day as part of our natural intake of foodstuffs. Nitrate taken into our bodies is excreted without being metabolised and it is only when nitrate is converted to nitrite, as by contaminated water supplies, that it can be harmful (especially to babies). A hypothesis was also put forward that such nitrite could combine with foodstuffs to produce nitrosamine, a suspected carcinogen that occurs in small amounts in smoked fish, beer and bacon. Studies of the subject have been inconclusive but it has been summarised in the following manner: 'The attempt to link gastric cancer incidence with fertiliser use does not seem to be justified as increasing use of the latter coincides with falling incidence of the former.'

The second environmental aspect of fertiliser usage relates not to their application to the land, but their removal from it and, to some degree, to their manufacture. And despite what is sometimes claimed, few fertilisers are immune from all of these lines of criticism. Fertilisers produced by straightforward chemistry in a factory clearly contribute to a despoiling, if not direct pollution, of the environment in the vicinity of that factory; the immediate surroundings of some of the large British fertiliser factories present as unappealing a vision as I know. Yet other fertilisers are mined, quarried or otherwise extracted from their place of natural origin and those who have seen the devastation of Pacific Ocean islands by rock phosphate removal or the despoiling of South American guano deposits might want to think twice before adding to the trade. But then I would guess that most of those gardeners who prefer to rely for their plants' nutrients on dried blood, bonemeal and hoof and horn have never seen the inside of a slaughterhouse. Ultimately, such matters are ones of individual choice and it is not for me to write on tablets of stone; however, critics of any one type of fertiliser should be careful to appraise the considerations relating to all of them before venting their spleen.

The aspect of the choice between organic and inorganic fertilisers that has always intrigued me most relates to their relative effects on the flavour of vegetables and other edible garden produce. Vegetables grown from organic fertilisers taste better, I am often told, despite the fact that plants actually make use of fertilisers not in the chemical form in which they are provided but after they have been converted to more basic components. Nitrogen to a plant is nitrogen, whether it originates in ammonium sulphate or dried blood. Taste is of course a subjective matter and if you believe you can detect a difference, all well and good; it may have some basis in scientific fact yet the evidence is strangely lacking. And in view of the multifarious chemical processes that go to make up that curious phenomenon called flavour, I am astonished, not so much that the type of fertiliser can affect it, but that an organic fertiliser should always change it in ways that we consider an improvement.

Whilst basic water solubility of its component chemicals is a major factor dictating whether a fertiliser is slow or fast acting, another factor can be superimposed upon this, its physical formulation. I have already mentioned in passing how grinding a substance finely will increase its surface area and hence its solubility, and, of course, the reverse is true. A particular compound or mixture of compounds can be amalgamated into a granule or other larger particle that presents a smaller surface to volume ratio and may also have incorporated within it a resin or other substance of lower water

solubility to regulate and slow down further the release of the active chemical principle.

Modern technology offers us, however, not only various formulations of the fertilisers themselves, but also the possibility of buying them in combination with other substances. Lawn fertilisers for spring and summer use, for instance, often have a weedkiller included, a significant labour-saving measure, whilst some of those for autumn and winter application are offered in combination with a mosskiller or (a development I deprecate), a worm killer such as carbaryl. But it is in combination with soil, peat or other growing media in seedling and potting composts that considerable strides have been made in fertiliser development in recent years.

The most difficult problem presented by the growing of plants in pots, or indeed anywhere other than in the natural soil of the garden, has been that of providing them with a satisfactory nutrient supply – although obtaining soils or other media with a physical constitution not prone to

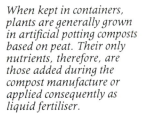

When kept in containers, plants are generally grown in artificial potting composts based on peat. Their only nutrients, therefore, are those added during the compost manufacture or applied consequently as liquid fertiliser.

compaction or capping has also been a significant complication. For soil-based growing media, the studies at the John Innes Horticultural Institution in the 1930s that led to the development of the John Innes seedling and potting composts laid down standards that we still adhere to today. Before that time, every commercial nurseryman and grower (and, to some extent, gardener too) had his own jealously protected formula for blending soil with manure, leaf mould, fertiliser and other ingredients. The major novel features of the John Innes range were their use of steam-sterilised loam of defined type, their use of peat instead of manures and composts, the accurate dosing of fertilisers and the discovery that a small range of potting composts would serve for a wide range of plants. In the original John Innes formulations, the fertilisers used were hoof and horn to supply nitrogen, super-phosphate for phosphorus and potassium sulphate for potassium, together with calcium carbonate; the blend being known as John Innes Base. Such composts were and are excellent when used fresh but problems were encountered with the mass-production and mass-marketing of them because of the use of hoof and horn as the nitrogen source. The release of nitrogen from hoof and horn by the action of fungi and bacteria begins as soon as the fertiliser is added to the soil and is influenced by the moisture content, pH and temperature of the compost. It is mediated through the production of ammonia. The change in pH that this causes, as well as the presence of the ammonia itself and the increase in salinity of the compost can cause serious problems to some types of plant. Genuine John Innes composts made to the formulae given should never be stored for more than two or three weeks therefore. Some modern commercial soil-based composts contain urea-formaldehyde or other artificial slow-release nitrogen source instead of hoof and horn and have a longer shelf-life although I would welcome date-stamping even of these. Standards of John Innes composts in Britain are checked through the John Innes Manufacturers' Association.

Formulations of John Innes Base fertiliser and John Innes seedling and potting composts

(a) John Innes Base fertiliser mixture

Fertiliser	Parts by weight
Hoof and horn	2
Superphosphate	2
Potassium sulphate	1
Calcium carbonate[a]	1

(b) John Innes potting composts

Ingredients		
		Parts by volume
Sterilised loam[b]		7
Peat[c]		3
Sand[d]		2

John Innes base fertiliser	kg/m³	(lb/yd³)
For potting compost No. 1	3.56	(6)
For potting compost No. 2	7.12	(12)
For potting compost No. 3	10.67	(18)

(c) John Innes seedling compost

Ingredients		
		Parts by volume
Sterilised loam[b]		2
Peat[c]		1
Sand[d]		1

Fertilisers	kg/m³	(lb/yd³)
Superphosphate	1.186	(2)
Calcium carbonate	0.593	(1)

[a] omit for calcifuge plants such as rhododendrons
[b] a medium clay loam of pH 5.5–6.5
[c] *Sphagnum* or sedge, pH 3.5–5.0, particle size 0.3–1.0 centimetres diameter
[d] coarse grit, 60–70 per cent of particles 0.15–0.5 centimetres diameter

After the Second World War, the increasing shortage of high-quality loam and increasing costs of transportation led to the development of alternative seedling and potting composts. Many materials have been used and continue to be appraised but far and away the most important for gardeners is peat, which is also the basis for the composts used in growing bags (p. 92). I have outlined the overall relative merits of peat-based and soil-based composts in Chapter Four, but it is important to appreciate that their fertiliser contents are rather different and the way that plants are fed in each is rather different too. Few if any gardeners are likely to go to the trouble of mixing their own peat-based composts from the individual ingredients although packs of ready blended fertilisers for adding to peat are readily available and offer a slightly cheaper option. A wide range of different N, P and K fertilisers is used in different branded products, most including modern slow-release nitrogen sources like urea-formaldehyde and with more minor nutrients in the form of frits or FTEs (fritted trace elements).These are in effect fragments of soft glass to which the nutrients have been bonded and from which they slowly dissolve. FTEs are manufactured by adding inorganic salts to sodium silicate, heating the mixture to around 1000 °C and then cooling it quickly.

Until recently, peat-based composts, in the same manner as John Innes composts, were divided simply into seedling (or seedling and cutting) and potting types. The former contained phosphorus to aid seed germination and early growth but little nitrogen and potassium. Now, however, the trend is to use varying amounts of the same fertilisers for the different composts and most modern seed-sowing and potting composts contain a slow-release compound fertiliser in mini-granular form, usually with FTEs. This fertiliser is based on magnesium ammonium phosphate with potassium sulphate; the most widely used of these fertilisers in Britain has the analysis 5:10.5:8.3 with 10 per cent magnesium. Typically, a compost sold for seed-sowing contains between 0.4 and 0.6 of the amount of slow-release fertiliser used for potting mixtures.

Ultimately, all plants grown in soil-less growing media will exhaust their nutrient supply, for the peat or other organic bulk has not nutrient reserve of its own. Additional feeding is essential and is almost invariably achieved by means of liquid fertiliser formulations such as those described on pp. 83–4. The technology of nutrient availability in soil-less composts has anguished both commercial growers and manufacturers alike and not all of the problems are entirely resolved. However, gardeners should not encounter difficulties if they follow the directions given by the compost and liquid fertiliser manufacturers. But because of the variation in fertiliser between composts of different brands, queries relating to specific nutritional problems that do arise are best directed to the manufacturer concerned.

I mentioned the relatively low nutrient content of organic manures and composts in Chapter Four and those gardeners who prefer to adopt 'organic' or 'natural' growing systems should always bear in mind that cropping imposes far greater demands on the soil than do wild plants in wild habitats. Seldom if ever is it possible to satisfy the nutrient needs of vegetables or, to a less serious extent, other garden plants, solely with organic manures without running into serious soil structural or other complications. A glance at the table above right will indicate the relatively low nutrient content even of the 'richest' manures such as chicken or pigeon. Note, for instance, that fresh pig manure contains on average 0.6 per cent by weight of nitrogen. The lower table shows that a summer cabbage crop needs about 25 grams per square metre of nitrogen during its life and if this is all to be supplied as pig manure, you will be faced with the pleasant task of unloading over 4 kilograms of it onto every square metre of your brassica bed; and even then, you will fall short of the amount of phosphorus and potassium required. Almost without exception, therefore, some fertiliser supplement will be needed, but the value of organic manures and composts as part of soil structure management is so high that their

Nutrient content of commonly used organic manures[a] and composts

	Approximate percentage contents of main nutrients			Noteworthy features
	N	P[b]	K[c]	
Manures				
Chicken	2.0	1.8	1.0	High in all nutrients but care needed to avoid N : K imbalance
Dog	0.4	0.2	0.1	Always compost kennel waste before use for health reasons
Pig	0.6	0.6	0.4	Strong smell a disadvantage other than in rural areas
Pigeon	3.4	1.4	1.2	Very high in nitrogen in relation to potassium
Rabbit	0.5	1.2	0.5	Usefully high in phosphorus but rarely available in large amounts
Sheep	0.8	0.5	0.4	Low in moisture; easy to handle but rarely available in large amounts
Stable	0.7	0.5	0.6	Lower moisture content than cow; easier to handle
Strawy cow (fresh)	1.2	0.4	0.5	Of greatest value as a soil conditioner in autumn
Strawy cow (old)[d]	0.4	0.2	0.6	Safe to apply in spring as less likely than fresh manure to deplete soil nitrogen
Composts				
Bracken	2.0	0.2	0.5	Better used as a mulch as it may deplete nitrogen as it rots
Garden compost	0.7	0.4	0.4	Care needed to avoid weed problems if inefficiently made
Leaf mould	0.4	0.2	0.3	Rots very slowly; better made separately from compost for mulching
Mushroom compost	0.6	0.5	0.9	High lime content; possible problems from insecticide residues
Sawdust	0.2	0.1	0.1	Not recommended; rots very slowly and may be toxic to plants
Seaweed	0.6	0.3	1.0	High in trace elements; attracts flies and best composted first
Silage waste	0.2	0.1	0.6	Only really useful for its fairly high potassium content
Soot	3.6	0.1	0.1	Compost first to avoid severe nutrient imbalance; possible toxic residues
Spent hops	1.1	0.3	0.1	Care needed to avoid N : K imbalance
Straw	0.5	0.2	0.9	Better used as a mulch as it may deplete nitrogen as it rots

[a] it must be appreciated that the attributes of manures will vary with the types of bedding used for animals and with other differences in the conditions in which they have been kept. Nutrient contents will also vary with the age of the material and the conditions in which it has been stored (see p. 80)
[b] as P_2O_5
[c] as K_2O
[d] stored under cover for six months

Approximate nitrogen requirements of common garden vegetable crops

Crop	Nitrogen requirement (grams per square metre)	Amount of Growmore required to satisfy nitrogen requirement (grams per square metre)
Cabbage	25	357
Brussels sprout	25	357
Cauliflower	19	272
Spinach	19	272
Beetroot	19	272
Potato (maincrop)	19	272
Potato (early)	15	214
Leek	15	214
Turnip	13	186
Beans (French)	13	186
Calabrese	13	186
Onion	11	157
Lettuce	11	157
Swede	9	129
Parsnip	9	129
Bean (broad)	9	129
Radish	2	29
Carrot	2	29
Pea	0	0

role as nutrient sources does merit close examination. Garden compost is obviously extremely variable and it is difficult to make specific comments about it so I shall confine myself to the second most widely used form of organic matter, farmyard manure, of which over 50 million tonnes annually are produced in Britain.

As they issue from the cow, dung and urine vary considerably in their relative contents of nitrogen, phosphorus and potassium, this variation depending on the way in which the cow itself has been fed and husbanded. And to become farmyard manure, the dung and urine are mixed with bedding material, usually straw, which is itself of variable composition. But thereafter, many other changes can occur before the material arrives in your garden, and if you hope to supply the optimum proportion of plant nutrient from manure, you should familiarise yourself with certain aspects of its handling before purchasing large quantities. Studies of strawy farmyard manure stored in a loose stack in the open over a single winter revealed that 10 per cent of the nitrogen was lost to the air – partly by partial decomposition to ammonia and partly by complete decomposition through denitrification to nitrogen gas. A further 10 per cent, together with 5 per cent of the phosphorus and 35 per cent of the potassium, was lost through seepage and leaching following rainfall. If, however, the manure has been compacted (through being trampled by the cattle, for instance), nitrogen fermentation tends to be anaerobic rather than aerobic, and very little nitrogen is lost. Thus, to retain as much as possible of the nutrients, manure should be stored in as compacted a form as possible, covered from the rain and stacked on a concrete or other water-retentive surface. As the application of fresh manure to soil is to be undertaken with caution (p. 55), and as farmers may be reluctant to store manure if they do not want it for their own use, the best plan is to buy fresh manure and store it carefully in your own garden or allotment until required.

Fertilisers can be divided into three groups depending on their relative contents of the three major elements, nitrogen, phosphorous and potassium and each of these groups is described below.

1. Nitrogen fertilisers

Ammonium sulphate (sulphate of ammonia) contains 21% N and is easily the most valuable and important fast acting source of nitrogen. Although it is fairly readily absorbed by organic matter and clay in the soil and thus partly prevented from being rapidly leached by rain, its ammonia component is rather quickly nitrified to form soluble nitrate which will soon be lost if plants are not present to make swift use of it. Thus, ammonium sulphate is slightly more persistent than straightforward nitrate salts and it is most commonly encountered by gardeners as the nitrogen donor to the popular general purpose compound fertiliser known as Growmore. A further feature of ammonium sulphate is that when added to soil, the sulphate component forms sulphuric acid, which in turn combines with any calcium present to produce soluble calcium sulphate. Thus, the fertiliser adds to the loss of calcium (actually to a loss equal to its own weight of calcium carbonate) and to an increase in soil acidity, a significant disadvantage on soils already acidic but a bonus when it is desirable to acidify an already alkaline one. Instead of ammonium sulphate, therefore, it is advisable to use the proprietary Nitro-Chalk in acid conditions. This contains ammonium nitrate and calcium carbonate. The ammonium nitrate contains 35% N, about half as immediately available nitrate with the remainder as slightly slower acting ammonia. It is granulated with the calcium carbonate partly for ease of handling but also because this addition balances loss of calcium brought about by ammonia.

Sodium nitrate (nitrate of soda) contains 16 per cent of immediately available nitrogen and was once the most widely used nitrogenous fertiliser. It has the advantage of not increasing soil acidity but is unlikely today to find any special use as a garden fertiliser.

In the principal organic sources of nitrogen for gardeners, the release of the nitrogen for plant use is not simply as a result of the materials dissolving in water but is dependent on enzymic decomposition by bacteria and fungi. The best-known material is dried blood which is understandably variable in composition but contains 10–13% N. The best-quality dried blood is produced by vacuum drying, lower grades by steam or dry heat. The availability of the nitrogen to plants is about 90 per cent of that in ammonium nitrate, most grades of dried blood being partially soluble in water. It is usually used as a component of blood, fish and bone (see p. 83). Fishmeal, containing 6–10% N, another component of this mixture, is also rarely used alone, but hoof and horn, with 7–16% N is often advocated as an organic source of slowly released nitrogen, the rapidity of the release varying with the fineness to which the raw material has been ground. The slowest release grades have a particle size of about 12 millimetres and are akin to coarse grit. Because of the mechanism of their breakdown, there are special problems associated with the use of all of these organic sources of nitrogen in potting composts (see p. 84).

2. Phosphorus fertilisers

The principal phosphorus-containing fertiliser is superphosphate, mainly a combination of mono-calcium phosphate and gypsum (calcium sulphate), formed by treating natural rock phosphate with sulphuric acid. It is the fastest acting phosphate source, containing 18 per cent of water-soluble P_2O_5 (see p. 73) and is used extensively in compound fertilisers. Although it supplies phosphate in an available water-soluble form, it is not rapidly lost from the soil because the phosphate slowly combines with other compounds, although remaining accessible to plants for a long time. However, gradually it becomes less and less useful to plants, especially in acid soils. In gardens other sources of phosphorus are also important. The ground residue of animal bones, after removal of more or less all of the gelatine and fats, contains calcium phosphate and is available either as bone meal, with 20–24% P_2O_5 and 3–4% N or as steamed bone meal or flour with 27.5% P_2O_5 and 0.8% N. The bulk of the nitrogen component is lost therefore in the steaming process. The phosphorus in bone flour, which is more finely ground, is more immediately available than that in bone meal. The phosphates in bone meal, like those in superphosphate, are not water soluble and are slowly made available to plants. Overall, neither superphosphate nor bone meal change the soil pH significantly and although it was formerly claimed that superphosphate acidified the soil by removing calcium, greater understanding of the soil chemistry involved now suggests that any effect is more likely to be a small one in the opposite direction. Nonetheless, the slight addition of calcium seems to have little deleterious effect on such calcifuge plants as rhododendrons, and bone meal may be used quite safely when planting them.

Rock phosphate itself, principally from North Africa, is available finely ground as a fertiliser and some gardeners find it more appealing in this form than when it has been converted to superphosphate. The composition is variable and ranges from about 25 to 35% P_2O_5, but even when very finely ground its solubility is such that the phosphorus is less available to plants than that from other materials. Extensive tests have indicated that rock phosphate is of most value on acid soils in areas of high rainfall. It is likely to be of very little benefit as a short-term treatment for annual plants.

Basic slag is another phosphorus-containing fertiliser, formerly more readily obtained and extensively used by gardeners than it is today. It is a by-product of steel manufacture, formed when lime is added to the blast furnace to remove phosphorus from pig iron, and it is finely ground for horticultural use. Basic slags are very variable substances and their value as fertilisers depends on their total P_2O_5 content and also on the proportion of the P_2O_5 that is soluble in a standard test involving citric acid. The so-called high-soluble slags are to be preferred but should be used cautiously by gardeners. They are traditionally farm fertilisers for grassland and some gardeners have interpreted this as suggesting they are useful on lawns. But their great merit on pastures is that they encourage the development of white clover; scarcely desirable on a lawn. The only justifiable use for basic slag in a modern garden is probably for pea and bean crops which, being related to clover, benefit similarly. Application of a high-soluble slag at about 140 grams per square metre in the preparation of a runner bean trench, for instance, may be worthwhile therefore.

3. Potassium fertilisers

There are only two potassium fertilisers widely used today and of interest to gardeners: potassium sulphate (sulphate of potash), which contains 48% K_2O and potassium chloride (chloride or muriate of potash), of which the high grade containing 60% K_2O is the most commonly encountered. Even high-grade potassium chloride contains a small amount of sodium chloride (common salt) as an impurity (lower grades contain much higher amounts of common salt) and this renders it attractive for farmers with such crops as sugar beet which respond favourably to sodium-chloride-containing fertilisers. In gardens, there is little justification for using the chloride salt because several salad vegetables, soft fruit and, to some extent, potatoes can suffer damage from chlorides. Potassium sulphate should be the main inorganic source of potassium, therefore, and it is generally applied to flowering and fruiting crops early in the growing season in the solid form in combination with nitrogen and phosphorus-containing products. Alternatively, it may be used alone to help stimulate flowering (on reluctant fruit trees and bushes for instance) at rates of about 35–70 grams per square metre.

As I have mentioned, some artificial compound fertilisers (especially liquid formulations) and many organic products contain small amounts of minor nutrients. In most gardening situations, I believe their merits are overstated (see p. 73) although they are important with soil-less growing media containing no natural nutrient reserves. In certain circumstances, however, fertilisers formulated specifically for their minor nutrient content are valuable. The most important are those containing sequestered or chelated iron which supply the element in a form that plants can assimilate readily in alkaline condition (see pp. 66–7). These should be used routinely on most soils at the start of the growing season for plants like hydrangeas, raspberries, roses, strawberries and other plants that are particularly responsive to iron shortage. Rhododendrons, azaleas, camellias and other strongly calcifuge plants also respond very favourably to sequestered iron in all except the most acid soils. The specific fertilisers needed to correct occasional deficiencies of boron, molybdenum and other trace elements are listed in the table on p. 68.

From a reading of the fertiliser details above, it should be possible to obtain an indication of the substances that you will require for any particular gardening task. But my experience is that most gardeners (and most good gardeners) manage perfectly well with only a short selection of the total range and my suggestions for the fertiliser requirements of an average home garden are listed below:

1. A balanced, general purpose solid fertiliser

The choice lies principally between an artificial granular fertiliser of the Growmore type and an organic mixture such as the dried blood, fish meal and bone meal blend called blood, fish and bone. Growmore types of fertiliser were originally called National Growmore mixtures and were introduced in Britain shortly after the Second World War. They took their name from the Growmore leaflets published by the Ministry of Agriculture during the war to help improve home vegetable production. A

When plants are reluctant to bloom, flowering can often be induced by feeding them with liquid tomato fertiliser or a similar compound liquid fertiliser with a high potash content.

Growmore fertiliser contains 7 per cent by weight each of nitrogen, phosphate and potash – a balanced 7:7:7 compound therefore – and although so-called liquid Growmore fertilisers are now available, the granular formulations are particularly valuable for use among vegetables and other plants at or just before the start of the growing season. The table on p. 79 includes the Growmore fertiliser requirements of the most common vegetable crops.

The principal organically based alternative to Growmore, blood, fish and bone is, like all organic fertilisers, of more variable composition but is approximately 5.1:5:6.5. The nitrogen from the dried blood tends to be slightly more slowly available than that from the ammonium sulphate in Growmore and the phosphorus is released more slowly from the bone meal component. Potassium content in pure blood, fish and bone is almost nil and so sulphate of potash is usually added to the proprietary mixture – hence the designation as 'organically based' rather than 'organic'.

2. A general purpose liquid fertiliser, relatively high in potassium

Bearing in mind the usefulness of a liquid feed during rapid summer growth and also the value of potassium for flower and fruit development, a fertiliser of this type should be the mainstay of most gardeners' fertiliser usage during the height of the season. There are several branded liquid products of this nature, varying in their relative nitrogen, phosphorus and potassium contents and virtually all containing additional minor elements – an advantage since they tend to be used extensively for plants growing in peat-based media as well as for those in soil. The concentrated liquid tomato fertilisers, for instance, generally with a composition of around 5:5:9 derived from inorganic components, are of this type. Most of the liquid fertilisers purchased in the form of soluble powders or crystals also fall into this category, having NPK ratios of about 1 J:5:20.

None of those liquid fertilisers I have mentioned so far are organically based and not all liquid fertilisers are rich in potassium. Indeed, the organic liquid fertilisers such as those based on seaweed extract generally have a low nutrient content overall (and thus larger volumes of them will be needed) and are noticeably low in potassium. Liquid manure, obtained mainly from cattle urine, is certainly an organically derived substance and contains both nitrogen and potassium but has almost no phosphorus (composition about 0.3:0:0.5). However, it is not readily available to gardeners and branded products with similar names may not be organically based and should not be confused with the genuine article.

3. Bone meal or superphosphate

I have described the chemical content and horticultural merit of superphosphate and bone meal on p. 81 and I consider there is little difference between them and no effective alternatives for the slow release of phosphorus to aid the establishment of perennials. They should be used routinely at the rate of about 175 grams per square metre in the planting of trees, shrubs, herbaceous perennials and bulbs therefore.

4. Two lawn fertilisers

Lawns should generally be fed twice a year, in spring and autumn, but the nutrient requirements of plants are different at these times. In spring and summer, the need is primarily for nitrogen to encourage strong and attractive dark green, leafy growth. This is best provided by a powder formulation of a nitrogen-rich fertiliser, preferably with at least some of the nitrogen in a slow-release form to extend the benefit for several weeks. Most branded spring and summer lawn fertiliser products have a major nutrient content of about 14:3:7 and should be applied at the rate of about 35 grams per square metre.

However, use of a high nitrogen fertiliser after the beginning of September is likely to encourage soft growth that will be prone to damage from winter cold and possibly from fungal diseases too. At this time, therefore, a powder formulation is needed of a fertiliser low in nitrogen but high in potassium (to stimulate the formation of harder leaf tissues) and also high in slow-release phosphorus (to encourage root development as growth recommences early in the new season). Generally, the analysis of such products is about 6:10:18 and they, too, should be applied at about 35 grams per square metre. Such a fertiliser is also valuable as a pre-sowing or pre-turfing treatment with new lawns, for although specially formulated fertilisers high in phosphate are used by professionals for this purpose, they are not generally available to amateur gardeners.

5. John Innes Base

I discussed John Innes Base fertiliser on p. 77 in relation to its principal role in providing a balanced supply of nutrients to loam-based potting composts. In this respect, it is of course an essential product for any gardeners mixing their own potting media but I find it valuable for a more widespread gardening purpose too. Most gardens contain at least some pots or other containers for summer flowers, bulbs, wallflowers and other ornamentals. Whilst it is advisable and feasible to replace the peat or loam-based compost in small pots and window-boxes every season, this is scarcely practicable with larger tubs and half-barrel-sized containers. But, if these

The period when growth commences in the spring is the ideal time to apply a general-purpose solid fertiliser to beds of herbaceous plants. The slow-release components will gradually make their nutrients available over the coming weeks.

are filled initially with a John Innes No. 3 potting compost, they can serve reliably for several years without being refilled if the top few centimetres of old compost are removed and replaced with a mixture of peat and John Innes Base (about 70 grams per square metre) at each replanting.

6. Rose fertiliser

The demands imposed by gardeners on their roses and the demands that the plants in turn impose on the soil's nutrient reserves are high. Accordingly, fertilisers have been formulated specifically for the purpose of feeding roses and these contain a blend of nitrogen, phosphorus and potassium, with special emphasis on the potassium to encourage flower development. Most branded products also contain additional magnesium (sometimes as an inorganic slow-release formulation) for roses are particularly prone to deficiency of this element. In general, I apply rose fertilisers at the rate of about 35 grams per square metre following the spring pruning and again after the first flush of summer flowers in late June. Although formulated spec- ifically for roses, such fertilisers provide an ideal balanced fertiliser for other flowering shrubs also and I feed them all at the same time.

Whether fertilisers are applied as liquids, granules or powders, it is important that they are spread as uniformly as possible in the area where they are needed. Powders and granules applied to individual plants or to rows of vegetables are almost always spread by hand and, with practice, it is fairly easy to obtain even coverage. Nonetheless, dosage should be judged carefully and handfuls of fertiliser not thrown around indiscriminately. Modern fertiliser packet labels generally bear dosing instructions in grams per square metre (and sometimes in ounces per square yard too). It is unreasonable to expect gardeners to weigh out the products every time they are used, but it is most important when applying materials by the handful to know approxi-

mately how much each of your handfuls contains. Not only do some gardeners have large hands and some small ones, but any handful of granular Growmore will weigh less than the same handful of bone meal. With the aid of a set of scales therefore, determine the weight of one of your own handfuls of each fertiliser you use and write this either on the packet or on a card pinned up in your garden shed. And always remember to wear gloves if you have cuts on your skin and always wash your hands after handling any fertiliser. It is worth commenting in particular on the special position of bone meal and bone meal-containing mixtures in relation to safety. There has for a long time been a slight risk of the bacteria responsible for causing anthrax being present in bone fertilisers and although this risk is almost negligible with most bone meal imported into Britain, precautions are nevertheless wise.

Lawn fertilisers in powder form can be applied by hand but even though most contain a coloured dye to render visible those parts already treated, it is very difficult to spread them evenly – and remember that inaccuracies will be revealed by a mosaic of patches of light and dark grass which persist for many weeks. Small-wheeled lawn fertiliser spreaders are relatively inexpensive and render uniform dosing much easier and quicker. It is wise to buy a spreader that can be calibrated variously so that it may be used with products from different manufacturers that may need slightly differing dose rates.

On a small scale, liquid fertilisers can be applied by sprayer or, especially on a lawn, by watering can, but for large areas of garden, two other possibilities are available. A hose-end diluter is a container that can be fitted to the delivery end of a hose-pipe. The container is filled with a concentrated fertiliser solution and the flow of water through the hose draws out concentrate to deliver diluted liquid feed. It is important to know the dilution factor used by the diluter for not all are variable and in such cases it will be necessary to vary the concentration of the liquid in the diluter when using different types of liquid fertiliser or

using them for different purposes. Never use such a diluter for any pesticide or weedkiller application however.

An alternative diluter delivery system is offered by some manufacturers who produce fertilisers of different types in the form of pellets which are fitted into a gun-like attachment on the end of a hose-pipe. The pellet is dissolved at pre-determined rate as the water flows over it. Such a system restricts you to pelleted fertilisers, however, and can be a more expensive option than either the watering-can dilutor or hose-end dilutor.

Whilst it might seem that spraying fertiliser over plants or placing granules or powders as close as possible to them would ensure the most efficient uptake, this is not necessarily so. Solid fertilisers applied in advance of sowing can be damaging to young seedlings although this effect can be minimised by watering the seed bed thoroughly after the fertiliser has been spread. Nonetheless, it is sensible to apply about a third of the recommended dose before sowing, water this in well, apply a further third after the young plants have emerged and the remainder when the plants are about a quarter grown. It is also possible to lessen further the possible damage to plants from fertilisers. I mentioned on p. 82 that some plants, even when mature, are sensitive to high levels of chloride in the soil and suggested this as a reason for avoiding the use of potassium chloride in gardens. In practice, however, high levels both of chlorides and nitrates can cause fairly general damage to young roots. This can be lessened by the technique of applying specific nitrogen and potassium-containing fertilisers between the rows of plants so that the high initial salt concentrations formed as the raw fertilisers dissolve are as far as practicable from the young root systems. However, if superphosphate or bone meal are being used, a rather different technique applies for these are not damaging to roots and there is, indeed, a positive benefit to emergence and early growth if these fertilisers are placed close to the plants. In a seed bed, therefore, try taking out an additional drill, about 5 centimetres to one side of the seeds and to a depth of about 5 centimetres below the seeds and place the appropriate dose of phosphorus fertiliser in this as a band. Not only does this ensure that the fertiliser is close to the seeds and the young roots, but the concentration of the product in the band means that it is absorbed more slowly into the soil. Among the vegetables that respond especially favourably to this treatment are French beans, carrots, lettuces and onions.

I have left until the end of this chapter the subject of liming, an activity which is subverted by mis-understanding and which impinges both on plant nutrition and soil management. Lime is a fairly pure form of calcium carbonate and the natural lime content of a soil not only governs the amount of the element calcium available for plants, but is also the major factor dictating soil pH and thus has major implications for the availability of other elements (see diagram on p. 67). It is in fact in the latter role that lime has its greatest horticultural importance for, given that a soil pH of about 6.5 (see pp. 41–2) is of greatest value for the widest range of plants (and especially of vegetables), clearly this is the value towards which gardeners should aim.

There is also, however, a widespread view among older horticulturists that lime has an appreciable effect on soil structure, in particular that it renders heavy clay soils less sticky. This has its basis in the well-known fact that a suspension of clay particles in water can be flocculated (formed into a woolly mass) by the addition of calcium salts. Whilst there may be some flocculation of soil colloids by the addition of lime, this in itself will not bring about crumb formation for the particles must still be bound together. And whilst lime may help break down the size of clods on some types of clay soil by direct physical chemistry, there remains the enigma of certain naturally limey clays that are consistently extremely difficult to work. It may be that much of the benefit believed to derive from the liming of acid clays is actually mediated through the improvement of conditions for bacterial develop-ment. The bacteria thus increase the breakdown of organic matter and it is this, in turn, that aids crumb formation.

However, lime should never be added to garden soil until a pH test (p. 169) has clearly demon-strated that it is necessary. And even when pH readings consistently below 6.5 are obtained, lime should only be added in amounts that take account of soil texture too. I have indicated on the right the amounts of lime as ground limestone that should be added to soils of various types in order to raise the pH to 6.5. In practice, lime is a generic term applied to several different substances and embraces not only limestone itself (calcium carbo-nate), but also quicklime (calcium oxide), slaked lime (calcium hydroxide) and the magnesian lime-stones in which the place of some of the calcium is taken by magnesium (legally, a magnesian lime-stone must contain at least 10 per cent magnesium carbonate). Because both slaked lime and quick-lime contain relatively larger amounts of calcium and have a higher neutralising value (they are more efficient at decreasing the soil acidity), they should be applied at half the rates given on the chart for ground limestone. For general garden use, however, ground limestone (often sold simply as garden lime) is to be preferred. It is cheaper in relation to its neutralising value, is unlikely to scorch vegetation, can be stored more or less inde-finitely and is much more pleasant to handle than the alternatives. Containing relatively less available calcium and being applied therefore in larger amounts, the spreading of limestone uniformly is also easier. And only when there is known to be a deficiency of magnesium in the soil should magne-sian limestone be used.

Given that a soil pH test indicates liming to be necessary, there are certain further considerations to take into account. Timing of lime application is important because the effects are not quick to appear. It is usually best to apply lime to land after autumn digging, and if a rotavator is available this can be used with advantage to turn the lime under – the operation is much too laborious to contem-plate by hand. Whilst limestone can be applied

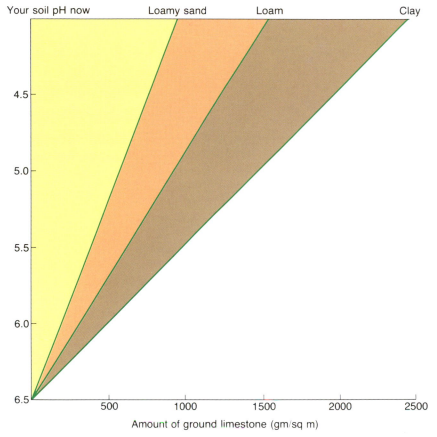

Your soil pH now — Loamy sand — Loam — Clay

Amount of ground limestone (gm/sq m)

Quantities of ground limestone to add to garden soil of various types in order to raise the pH to 6.5 (the value of most general use) from different starting levels.

closer to sowing or planting time, this is generally inadvisable and certainly neither hydrated nor quicklime should be applied in this way for they will almost certainly cause scorching of the young plants. There is almost never a need for annual liming – the maximum effect on soil pH does not occur until about four years after application. The losses of lime are usually greatest from light soils with a low organic matter content, and on this type of land a lime application every three or four years should be given – provided always that a pH test indicates it to be desirable. On heavy soils, once every six or seven years will almost certainly suffice. And lime should never be applied at the same time as animal manures for it will react with them to liberate ammonia; an absolute minimum period of one month should elapse therefore after liming and before manure is applied.

The gardening effort required in keeping a lawn well-fed and maintained is rewarded when the result is a classic 'bowling-green' surface.

CHAPTER SIX

Plant husbandry

I can define the subject of this chapter quite simply as the major differences between cultivation and natural plant growth. However, cultivation is manifest in ways both many and varied, some of which differ much more than others from natural behaviour. I have selected the most important components of the subject of plant husbandry, therefore, and shall discuss the principles behind each and the ways in which our gardening procedures capitalise on the attributes and capabilities of plants that nature has provided.

'Growing System' is a useful expression to describe the overall features of the environment within which plants are cultivated. The most familiar garden examples of growing systems are the open ground, the various protected systems such as cloches, frames and greenhouses, and contained or miniaturised systems such as pots, window-boxes, hanging baskets and growing bags. There is overlap between some of them (the use of growing bags in greenhouses, for instance), and there are a number of overriding features that apply to several. These overriding features include the significance of monocultures (repeated growing of one species only) vis-à-vis mixed cropping and of the variations in plant to plant spacing.

Plants grow naturally outdoors, and it might be thought therefore that growing them in beds and borders in the open garden comes fairly close to mimicking nature. There are nonetheless several significant differences. The choice of plants is that of the garden owner and not the result of the competitive struggles between species thrown together by the chances and vagaries of the conditions offered by the garden, they will certainly be grown with companions that they would never encounter in the wild state, and they will almost certainly be grown apart from others that are their natural bedfellows. The disposition of the individual plants will be very probably on some more or less regular geometric pattern (one obvious extreme is the serried ranks of the vegetable garden), but even within such an apparently heterogeneous assemblage as the herbaceous border, each species will almost certainly be planted in discrete groups of three or five individuals and the groups themselves will be dispersed predictably among groups of other individuals of other species. But not only will the spatial arrangements of the plants differ from those occurring naturally; their sequential relationships in time will vary too. Cabbage and carrot plants are not wrenched naturally from the ground and taken elsewhere at the peak of their maturity, to be replaced rather swiftly afterwards by other plants, often of the same or related type. Yet in gardens, this is precisely what happens every year.

The protected environment offers no pretence at resembling anything natural for it is simply a device to enable plants to be grown in climates harsher than those in which they occur naturally, or to prolong their productive life at the start and end of the outdoor growing season. But in protecting plants, we protect pests and disease-causing organisms too, often with serious consequences (ch. 8). Nor should we delude ourselves into thinking that all plants require the same type of protection from the same environmental undesirables. The subtropical cucumber in its high demand for warmth and humidity will not flourish readily alongside a warm but drier-climate tomato or grapevine and will certainly have very different needs from the alpine that is to be protected against extremes of heat and hanging dampness rather than cold. The

Whilst a large herbaceous border might superficially have much in common with a natural habitat, the plants are arranged in more discrete groups and actually comprise species that would be unlikely to grow together naturally.

ways in which a compromise may be effected are discussed in Chapter Nine.

The miniature or contained growing system has attributes and problems all of its own. There is a finite supply of nutrient, water and, of course, space. Regular and careful attention to supplementary feeding and watering are essential therefore, and although weeds are unlikely to present serious problems, pests and diseases may well thrive in the closeted atmosphere. It is easy both to under- and over-feed and water plants growing in any type of container, but growing bags are perhaps the hardest to manage effectively. The plants they contain (usually tomatoes) have a high demand for water and fertiliser in the height of summer yet, with little exposed compost surface, it is often very difficult to determine how moist or dry the contents of the bag have become. Container gardening embraces, moreover, two of the most extreme and bizarre environments within which we expect our plants to flourish. The hanging basket (a ludicrous idea when thought of objectively) is one; the ring culture system (an extremely clever one) is the other.

Hanging baskets suffer from several fairly obvious problems. The compost in them is usually peat rather than soil-based because of the problems of weight that the latter imposes. But like all peat-based composts, nutrient depletion is a serious difficulty, one compounded in this instance because of the very shallow container and the extremely frequent watering that it requires. This regular through-flow of water leaches nutrient very rapidly and it is almost necessary to add liquid fertiliser with every watering. The ring culture method is a valuable technique for growing plants (tomatoes especially) in greenhouses. It entails growing the plants in a soil-based compost in a bottomless pot or ring. This is placed on a bed of gravel, approximately 45 centimetres deep, in a trench lined with plastic sheet. The principle depends on the plant developing two root systems: one in the gravel used for water uptake and one in the compost to which liquid fertiliser is applied.

The hanging basket must be the most bizarre plant habitat ever devised: it is wholly dependent for its success on constant attention to feeding and watering.

Many of the component features of growing systems that I have mentioned will be touched upon in other chapters, but there are certain aspects of plant husbandry best discussed separately. First are the implications of the repeated growing of plants on the same area of soil and the removal of them after a period of time that falls short of their natural lifespan; many vegetables (carrots for instance) are biennials but are only allowed to grow in gardens for under twelve months. Moreover, the plants themselves have been selectively bred in various ways that almost invariably result in them having a higher nutrient demand on the soil than that exercised by their wild relatives. There are several consequences of this. The nutrient content of the soil will be depleted rapidly by a combination of the abnormally high demands of each plant and by the almost total lack of any natural return of plant debris to the site. Moreover, whilst any monoculture facilitates the build-up of pest and pathogen populations (p. 146), a repeated monoculture of the same type gives particular opportunity for the establishment of perpetuating populations of the many seriously damaging soil-inhabiting organisms. To some extent, pests and pathogens can be combated with insecticides and fungicides, and nutrient depletion can be corrected with fertilisers. Both of these procedures are wisely kept to a minimum however, and maximum use should be made of one purely cultural expedient to lessen difficulties. That expedient is usually called crop rotation, and it is in the vegetable garden that it achieves greatest success.

The principles of crop rotation are basic enough; they are simply to ensure the greatest possible time interval between successive crops of the same type of plant on the same area of soil. Thus, the differing nutrient requirements of different types of plant will ensure that a greater part of the soil nutrient spectrum will be utilised – grow a plant with a high potassium demand, for instance, this year and one with a high phosphorus demand next year, to put it simplistically. Moreover, given the break between

plants of the same type and given the fact that many pests and pathogens are fairly specific in their plant hosts, the populations of these organisms in the soil should die away between crops. There are in practice, however, serious shortcomings in the pest-and disease-controlling benefits to accrue from crop rotation in a garden, and I shall discuss these fully in Chapter Eight. The nutritional advantages are more evident, but a further and less appreciated benefit from crop rotation lies in its role as a contribution to overall soil management. The soil for different crops is prepared and cultivated in different ways, and a rotation scheme should ensure that shallow cultivations and possible consequent pan problems (p. 51) are not restricted permanently to one area.

In planning any vegetable crop rotation, certain information about the nutrition and general cultivation of the various plants is needed. A glance at the table on p. 79 will indicate the very widely varying nitrogen demands of different vegetables. Because nitrogen is the most important single element for plant growth (p. 69), it is largely on differing nitrogen requirements that cropping plans are designed, although it must be remembered that there are also variations between plants in their requirements for all other mineral nutrients. The table also indicates that the average vegetable garden contains around twenty different types of crop; clearly a crop rotation scheme to ensure that none of them was grown on the same area of soil for two or even more years in succession would tax gardening mathematics to the full. It would almost certainly also require that similar areas be occupied by each type of crop and that each occupied the ground for similar lengths of time; clearly a practical nonsense. A compromise is needed.

Although vegetables do differ in their nutrient needs, the major crops can fortunately be placed in three groups. These groups generally reflect the broad nutrient requirements of the vegetables and also certain other aspects of plot cultivation. The first group comprises primarily leafy plants with a very high nitrogen requirement – spinach and leafy

The vegetable garden benefits greatly from the process of rotation. For instance, keeping brassica plants together as a group is one key to successful cultivation.

brassicas such as Brussels sprouts, cabbage, cauliflower, kale, broccoli and calabrese (although the last two actually have a lower nitrogen demand than most of the others). To them are usually added swedes, turnips and kohl-rabi, which have much lower nitrogen needs and differ also in being root rather than leaf crops. The reason they are included here is because of their susceptibility to the soil-inhabiting clubroot organism and the treatment to combat this is considered to override all other considerations. I shall return to this important matter again later (p. 146).

The second group includes so-called root crops with a generally intermediate nitrogen requirement – carrots, beetroot, parsnips, onions and potatoes (although the latter two are, of course, not strictly root, but swollen stem crops). These plants are traditionally grouped together because of a tendency displayed by some of them to grow poorly or actually to be damaged on soil recently treated with fresh manure. The reasons for this effect are not entirely clear (although the temporary high concentration of organic acids seems to be in part responsible), but it should be appreciated that it applies only to fresh manure. Well-rotted manure or compost can be applied with advantage before sowing or planting all root crops, which benefit from the organic amendment. Indeed, carrots, which are the crop most adversely responsive to fresh manure, are grown commercially on highly organic soils and, in any event, there are other reasons such as stoniness or pan formation below the surface that can result in the forked or fanged roots usually attributed to manure damage.

The third group comprises most of the remaining vegetables, including the cucumber family and, very importantly, peas and beans. These last two have a very low nitrogen fertiliser requirement because of their ability to obtain nitrogen directly from the air through the medium of the bacteria-containing nodules on their roots (p. 43).

Not all of the common vegetables are included in

A three-course rotation scheme for a full vegetable garden. Clearly, many gardens do not have the space, nor gardeners the inclination to grow this much produce but the system can be adapted to individual requirements. It is important always to try and keep the three major categories of root crops, pod crops and brassicas separate.

these three groups of course; sweet corn, rhubarb, celery, herbs, tomatoes, lettuces, radishes and other salad plants are obvious exceptions. This is generally because, like rhubarb and herbs, they are grown semi-perennially on the same site or because, like radishes, they are catch crops, that is very fast-growing plants with few specific site requirements that can be sown and harvested between other crops or after some of the major vegetables have been removed. With all of these

considerations and constraints in mind, a possible three-plot rotation scheme for a home vegetable garden is shown below. This offers self-sufficiency in virtually all vegetables, but to facilitate amendments to suit individual gardeners' own requirements and tastes, I have also included overleaf a listing of the approximate yields per unit area to be expected for the common garden vegetable and fruit plants.

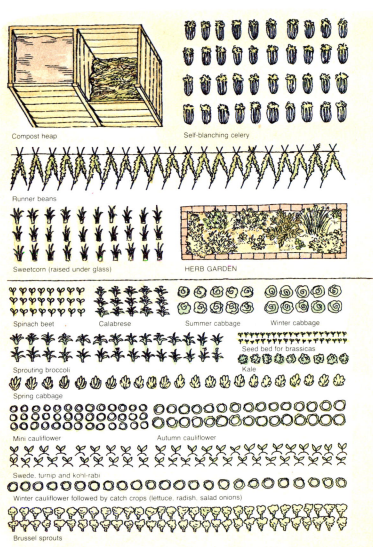

Compost heap

Self-blanching celery

Runner beans

Sweetcorn (raised under glass)

HERB GARDEN

Spinach beet

Calabrese

Summer cabbage

Winter cabbage

Seed bed for brassicas

Sprouting broccoli

Kale

Spring cabbage

Mini cauliflower

Autumn cauliflower

Swede, turnip and kohl-rabi

Winter cauliflower followed by catch crops (lettuce, radish, salad onions)

Brussel sprouts

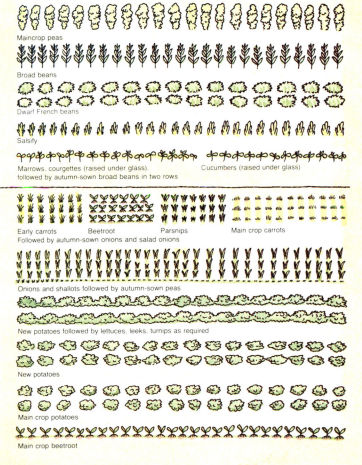

Tomatoes

Leeks

First early peas

Second early peas

Maincrop peas

Broad beans

Dwarf French beans

Salsify

Marrows, courgettes (raised under glass), followed by autumn-sown broad beans in two rows

Cucumbers (raised under glass)

Early carrots

Beetroot

Parsnips

Main crop carrots

Followed by autumn-sown onions and salad onions

Onions and shallots followed by autumn-sown peas

New potatoes followed by lettuces, leeks, turnips as required

New potatoes

Main crop potatoes

Main crop beetroot

Approximate average yields per unit area to be expected with some common garden soft fruit and vegetable crops[a]

Crop	Yield in kilograms (per 5 metres of row unless otherwise stated)
Soft fruit	
Black currant	5–7/bush
Gooseberry	4–6/bush or 1.5–2.5/single cordon
Loganberry and similar hybrids	4–7.5
Raspberry (autumn)	1–2.5
Raspberry (summer)	5–7.5
Red currant	4–5/bush or 1–1.5/single cordon
Strawberry	0.5–1.5
White currant	4–5/bush or 1–1.5/single cordon
Vegetables	
Bean (broad)	15[b]
Bean (French)	8
Bean (runner)	30–40
Beetroot	10–12
Broccoli	7–10
Brussels sprout	10–12
Cabbage (spring)	15–20
Cabbage (summer)	12–16
Cabbage (winter and savoy)	15–20
Calabrese	12–20
Carrot	13–26
Cauliflower (autumn)	8[c]
Cauliflower (winter)	8[c]
Celery	8
Courgette	5–9
Cucumber (outdoor, ridge)	*
Cucumber (unheated greenhouse)	*
Kale (curly)	7–10
Kohl-rabi	5–7
Leek	10–15
Lettuce	15–20[c]
Onion (bulb)	8–15
Onion (salad)	**
Parsnip	30–35
Pea	8–15[b]
Potato (early)	7–10
Potato (maincrop)	10–13
Radish	**
Shallot	5–7
Spinach	***
Swede	7–10
Sweet corn	15–20[c]
Tomato (outdoor)	12–18
Tomato (unheated greenhouse)	25–35
Turnip	7–10

[a] of course, yields will vary not only with growing conditions (and skill) but also with cultivars
[b] in pods
[c] numbers of individual plants or fruits, not weight
* two plants should suffice for a family of four
** sow regularly through season as catch crop
*** cut and come again crop; a 10-metre row should suffice for a family of four

My last paragraph touched on two important aspects of garden husbandry; neither entirely specific to the vegetable plot. The use of fast-growing and maturing plants such as radishes or lettuces has obvious merits when attempting to obtain the maximum food output from a given area of land. But the concept has wider applications in the garden too. Fast-growing annual bedding plants may be used to advantage as temporary filling for gaps in a herbaceous border of larger and slower growing trees and shrubs, to be removed as the larger specimens mature. The table opposite lists useful fast-growing plants for different purposes.

However, a major consideration implicit in the devising of a crop rotation or any other garden planting scheme is that of the spatial separation of the plants. The spacing of garden plants, especially vegetables, has a historical basis, and only fairly recently has scientific study revealed that some of the time-honoured practices do not stand up to critical analysis.

First, a dip into history. Before mechanical appliances were available, farmers sowed their seed by broadcasting it – handfuls of seed were thrown out as they walked across the field. The procedure is still used in undeveloped and impoverished parts of the world, but its inefficiency is fairly evident – some of the seeds will fall in clumps and some will fall sparsely. Not only does this have consequences for the ease and efficiency with which the young plants grow, but the subsequent maturing and harvesting of the crop will be uneven. In the early eighteenth century, the English gentleman farmer, Jethro Tull, revolutionised the growing of cereal crops with his horse-drawn corn drill and inter-row cultivator. The former placed the seeds in regular rows and the latter enabled weed growth between the rows to be controlled. Crops grown with row seeding yielded much better than those from broadcast seeding and, in time, the technique passed from cereals to field vegetable cultivation and thence to gardens. Not until quite recently was any question raised to the deduction that the plants yielded better *because* they were grown in rows. In

A selection of useful fast-growing plants for different garden purposes[a]

Annual ornamentals for beds and for filling gaps in borders

Calendula (pot marigold)
Iberis (candytuft)
Lobelia
Schizanthus (poor man's orchid)
Tithonia (Mexican sunflower)

Climbing plants (with approximate spread per year (in metres), at least for early years)

Akebia quinata (1.2–1.5)
Clematis (large-flowered hybrids) (1.2–2.0)
Clematis montana (0.6–1.0)
Clematis tangutica (1.5–2.5)
Hedera helix cvs. (ivies) (0.6–1.0)
Humulus lupulus 'Aureus' (golden hop) (1.2–1.5) (annual)
Lathyrus (sweet pea) (1.0–1.5) (annual)
Parthenocissus tricuspidata (Boston ivy) (1.2–2.0)
Periploca graeca (silk vine) (1.2–1.7)
Polygonum baldschuanicum (Russian vine) (2.5–4.0)
Rosa filipes 'Kiftsgate' (3.0–4.0)
Rosa longicuspis (2.5–3.0)
Tropaeolum majus (climbing nasturtium) (1.8–2.0) (annual)

Herbaceous perennials (tall-growing and/or spreading)

Allium giganteum
Arundinaria anceps (bamboo)
Arundinaria japonica (bamboo)
Cortaderia selloana (pampas grass)
Delphinium especially 'Pacific Hybrids'
Echinops ritro (globe thistle)
Eremurus hybrids (foxtail lily)
Geranium spp. (selected cvs. especially *G. oxonianum* 'Claridge Druce')
Gunnera manicata
Miscanthus sacchariflorus (grass)
Nepeta spp. (catmint)
Phalaris arundinacea var. *picta* (Gardeners' garters – grass)

Low-growing perennial ground cover

Alchemilla mollis (lady's mantle)
Cotoneaster dammeri
Erica (selected cvs.) (heaths)
Hedera colchica (ivy)
Hedera helix cvs. (ivies)
Hypericum calycinum (rose of Sharon)
Rosa 'Nozomi'
Rosa sancta
Rosa 'Snow Carpet'
Salix repens
Vinca major

Shrubs (with approximate heights (in metres) on good soil after three and ten years)

Buddleia davidii cvs.	2.5–3.0
Cornus alba 'Sibirica'	1.5–2.5
Cotoneaster X 'Cornubia'	1.8–4.0
X *Cupressocyparis leylandii*	2.5–12.0
Elaeagnus pungens 'Maculata'	1.5–2.0
Forsythia intermedia 'Spectabilis'	2.5–3.0
Kolkwitzia amabilis 'Pink Cloud'	1.8–3.0
Philadelphus coronarius	2.5–3.0
Philadelphus 'Virginal'	1.8–3.0
Rosa 'Amy Robsart'	2.5–3.5
Rosa doncasterii	3.0–4.0
Rosa 'Frühlingsgold'	2.5–3.5
Rosa 'Til Eulenspiegel'	2.5–2.5
Salix gracilistyla 'Melanostachys'	2.0–3.0
Sambucus racemosa 'Plumosa Aurea'	1.8–3.0

Vegetables and herbs for intercropping and space filling

Borage
Fennel
Onion (salad)
Radish
Swiss chard

[a] but always remember that fast-growing can mean invasive and, with perennials, difficult to eradicate if later you change your mind

one of those scientific conclusions that seems so simple once you know, it has emerged that the plants grown in rows only yielded better because of the more efficient weed control that it facilitated; not because of anything innately special in the row itself. Indeed, when it is subject to rather more critical appraisal, row cropping (apart from the weed control advantage) is in many instances seen to be biologically inefficient.

It is evident from what I have mentioned several times here that plants compete with each other for nutrient, moisture, air and light. And a little thought will indicate that the relative closeness of two individual plants governs just how great that competition will be. Plant two oak trees five metres apart and, before many years are done, their roots will be intertwined and their branches interlocked. Plant them one metre apart and either or both will very soon show series distortion and shortfall in some aspect of their performance. Yet two radish plants placed with one metre between them could as easily be over opposite horizons for all the effect that their positioning will have on each other. Irrespective, moreover, of the absolute distances involved, plants grown in rows have two quite different spacings – that between individual plants within the row and the greater distance between the rows themselves, very often a no-man's land where weeds flourish. This, too, is scientific nonsense, and not only is it more logical to space plants equidistantly, but the offsetting of adjacent individuals makes optimum use of the ground area.

Obviously, for every type of plant, there is an optimum spacing distance below which the competition causes a diminution in growth, although there are two ways in which this optimum can be assessed. Optimum growth can be defined either as that resulting in each individual plant attaining the maximum size or as the maximum total yield of plants per unit area. The two do not necessarily coincide, and if you wish to obtain the greatest total weight of vegetables from your plot, you may have to accept individual plants of slightly smaller size than you really need.

Clearly, there can be a compromise and it is from a combination of practical desirability and scientifically determined absolute maxima that the data on the right have been produced. Nonetheless, the ability to use different spacings actually to regulate the size of your produce rather than this being merely a consequence of your action can be very helpful. The most useful applications of this approach arise with lettuces, cauliflowers and onions. Close spacing of some lettuce cultivars (especially some of the cos types such as 'Lobjoits') will produce heartless plants (so-called leaf lettuce) which may be useful for salad making where large quantities of open leaves are desirable. Some early summer cauliflowers, especially 'Dominant', produce miniature, individual-portion-sized curds when they are planted very closely together while the bulbs of many onion cultivars can range in size from giant show specimens to tiny pickling forms simply through choosing the appropriate spacing. I have included details of these variations in the table opposite.

For obvious reasons, such optima have only been determined by trial and experimentation for the most valuable food plants, but most good nursery catalogues suggest plant-to-plant spacings for commonly grown garden ornamentals, sometimes in the form of plant densities – the number of plants required per square metre. You should bear in mind, however, that these spacings relate to the closeness of plants of the same type. In an ornamental planting, there will almost certainly be a

The optimum use of a given area of plot is obtained if the individual plants are off-set one from the other. The gaps of wasted space between the lettuces on the right are larger than those on the left.

Suggested plant-to-plant spacings for a range of common garden soft fruit and vegetable crops

Crop	Spacing (centimetres)		Crop	Spacing (centimetres)	
	Between rows	Between plants		Between rows	Between plants
Soft fruit			Cauliflower (winter)	75	75
Black currant	150	150	Cauliflower (summer, miniature curds)	22	10
Gooseberry (bush)	150	150	Celery (self-blanching)	25	25
Gooseberry (single cordon)	150	30	Celery (trenched)	–	15
Gooseberry (double cordon)	150	45	Courgette	90	90
Loganberry and similar hybrids	200	250–300	Cucumber (outdoor, ridge)	90	90
Raspberry	175	40	Cucumber (unheated greenhouse)	–	45
Red/white currant (bush)	150	150	Kale (curly)	45	45
Red/white currant (single cordon)	150	30	Kohl-rabi	30	15
Red/white currant (double cordon)	150	45	Leek	30	15
Strawberry	75–90	40–45	Lettuce (hearting)	25	25
			Lettuce (leaf)	12	7
			Onion (large bulb)	25	15
Vegetables			Onion (small bulb/ pickling)	25	2
Beans (broad)	23	23	Onion (salad)	15	2
Bean (French)	45	23	Parsnip	30	15
Bean (runner – single rows)	60	15	Pea	12	12
Bean (runner – double rows)	30	13 (45 between each pair)			(45 between triples)
Beetroot (early)	18	12	Potato (early – large-sized[a] tubers)	60	30
Beetroot (maincrop)	30	15	Potato (maincrop – medium-sized[b] tubers)	75	30
Broccoli	45	45	Radish	10	3
Brussels sprout	75	75	Shallot	25	15
Cabbage (spring)	30	30	Spinach	30	15
Cabbage (summer)	45	45	Swede[c]	30	20
Cabbage (winter and savoy)	45	45	Sweet corn[c]	45	30
Calabrese	45	45	Tomato (outdoor)	45	45
Carrot	15	10	Tomato (unheated greenhouse)	–	35
Cauliflower (autumn)	55	55	Turnip	30	20

[a] less than 12 tubers per kilogram
[b] about 14 tubers per kilogram
[c] almost always succeed best in beds rather than isolated rows

mixture of different types of plant so you will need to calculate an average value in such cases. For instance, if the planting distance for paeonies is given as 45 centimetres and that for heleniums as 30 centimetres, a paeony and a helenium placed adjacent to each other should be at a distance of about 38 centimetres.

But returning again to the reason that Jethro Tull abandoned broadcast seeding in favour of rows, it will be evident that if we now move away from rows and space plants equidistantly, we are likely to inherit the old difficulties of weed control, for the inter-row paths will no longer be present for us to walk along. The answer is to do on a small scale as the modern commercial vegetable grower does in his fields – grow the plants within beds. In fields, the width of the beds is determined by the need for them to be straddled by tractor wheels, but in gardens the criterion is that weed control and other cultivations can be accomplished easily from either side. A bed width of 1.2 metres has proved itself the most convenient, certainly for weeding, although if this proves too much of a stretch for planting or sowing, a temporary bridge comprising a strong plank resting on bricks should solve the problem. There are soil management considerations in the use of the bed system of cultivation, but these are covered in Chapter Four.

The commercial use of beds within which plants are spaced equidistantly should be adapted for gardens, as it is indisputably a more efficient system than row cropping. In commercial horticulture, the actual width of the beds is governed by the need for the tractor to be able to straddle them. For gardeners, the main criterion is that individual rows be within comfortable reach from both sides.

One of the first discoveries that the newcomer to fruit and vegetable gardening makes is the relationship between glut and scarcity; the narrow line that exists between feast and famine. The gardener soon realises how the 'hungry gap' arose in olden days when the end of the storage life of last season's crop did not coincide with the beginning of the new season's harvest. One of the greatest arts in food crop production of any kind is to spread the maturity of the produce as evenly as possible across the year. One obvious way is by making optimum use of the potential offered by the various storage methods (pp. 109–12), but another is by making the correct choice of cultivars and planting times. Some cultivars are more cold or heat tolerant than others, either in germination or growth, and some subsequently grow and mature faster. These are all important features with any crop plant, but especially so with those vegetables that offer the possibility of raising more than one crop within twelve months. I have indicated opposite some possible blueprints for obtaining continuity of produce with those garden vegetables most adaptable to the technique.

There is much more to plant husbandry than simply growing a plant at the time and in the place that is convenient to you and providing it with the basic essentials of nutrients, water, light and air.

How to obtain continuity of produce by sequential sowing and careful selection of cultivars of cabbages, lettuces and peas[a]

Time of maturity	Time to sow	Time to transplant	Recommended cultivars
Cabbage/Greens			
March–May	End July	September	'April', 'Avoncrest', 'Harbinger', 'Offenham', 'Spring Hero'
May–June	End February (greenhouse)	April	'Golden Acre', 'Hispi'
July–September	End March (frame/cloche)	End May	'Hispi', 'Marner Allfruh', 'Winnigstadt'
September–December	End April	June	'Best of All' (savoy), 'Christmas Drumhead', 'Jupiter'
December–February	May	End June	'Avon Coronet', 'Celtic', 'January King' (savoy)
Lettuce			
June–October	February (cloche) – July[b]		'Little Gem' (cos), 'Tom Thumb' (butterhead), 'Webb's Wonderful' (crisphead)
October–December	August (cloche)		'Avoncrisp' (crisphead), 'Avondefiance' (butterhead)
January–March	August (cold greenhouse)		'Kwiek'
April–May	September (cloche)		'Arctic King' (butterhead), 'Valdor' (butterhead)
Peas			
End May–June	October (cloche)		'Feltham First', 'Meteor', 'Pilot'
June–July	February–March (cloche)		'Early Onward', 'Feltham First', 'Kelvedon Wonder', 'Meteor'
End June–September	March–May		'Hurst Beagle', 'Hurst Green Shaft', 'Onward', 'Senator',
September–October	June		'Kelvedon Wonder'

[a] this is an outline scheme only; details of the sowing and transplanting conditions can be obtained from seed catalogues or packets
[b] after April, sow a new row as the seedlings from the previous sowing emerge

Many a potentially good garden plant, treated in this way, will still fall far short of your expectations of size, shape, flower and/or fruit production if its natural inclinations to grow in a particular manner are not restrained. For many types of plant, all of the feeding and watering in the world will be almost as nought if careful attention is not paid to pruning and training. Yet these are perhaps the least understood of all the gardener's techniques. Far too few gardeners prune plants correctly, and even many of those who do fail to appreciate the underlying principles. There are several different reasons for pruning, but all depend on removing certain parts of a plant and, at the same time, encouraging other parts to grow.

Plants may be pruned simply to keep them within an allotted space or to a required shape, to remove dead or diseased shoots and encourage the growth of new healthy tissue, or to prevent overcrowding of branches and hence the encouragement of conditions within which diseases and pests may flourish. But whatever the motive, pruning is the removal of some parts of a plant at the expense of others. The first key to understanding pruning, therefore, is to appreciate the way in which plants grow (in other words, the way that they produce new tissues), and also the way in which they branch.

As I explained in Chapter Two, the tissue growth occurs from specialised regions called meristems

Although the branch cuts on this tree have been made correctly, fairly close to but not flush with the trunk, the application of a black sealing compound is a useless and possibly harmful treatment.

It is most important to follow the correct sequence of events if a branch is to be removed cleanly and with no damage to the trunk of the tree. Broken or torn tissues will provide entry points for decay fungi.

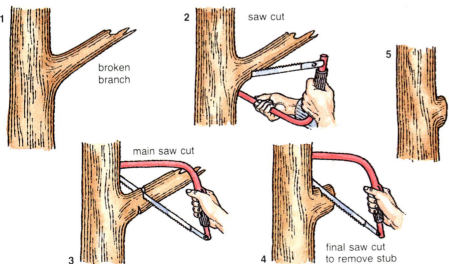

and, although new meristems can sometimes arise in almost any type of tissue and almost any part of a plant, those responsible for elongation of stems are normally localised at their tips in buds. And as I also explained in Chapter Two, stem branches arise from buds in leaf axils and from nowhere else. Thus, a pruning cut to anywhere other than immediately above such a bud will almost inevitably result in a length of stem that is unable to grow, generally unable to produce new meristems and new buds, and likely therefore to die and act merely as an entry point for pests and diseases. In saying this, of course, I am stating the ideal – any gardener asked to ensure that every shoot on his hedges was always trimmed back neatly to just above a bud would rightly dismiss both me and my advice as academic nonsense. But a sound general maxim is to bear in mind that the larger the plant and the larger the shoot being pruned, the more care should be exercised with the positioning of the cut. With large branches on trees and shrubs, special criteria apply, for the cut here is to the main stem or trunk and not to a bud.

Detailed studies in recent years, especially in the United States, have shed much new light on our understanding both of the healing processes in large woody tissues and on the manner in which decay-causing and other fungi invade them. The diagram indicates the sequence to follow when removing a branch or a jagged stub left after gale damage. It is most important not to cut into the swollen collar at the base of the branch (and certainly not to cut flush with the trunk as was once recommended), for the meristematic tissues responsible for forming wound callus and regenerating more permanent healing tissues are thus damaged. Nor is it desirable to treat wound surfaces with a healing compound which can actually inhibit natural healing, cause asphyxiation of the tissues and trap in any existing fungal growth.

Generally, a bud will produce a new shoot in the direction in which it points and, in most cases, it is desirable to induce growth outwards rather than inwards to minimise branch congestion. The general rule with roses and most flowering shrubs, therefore, is to prune to an outward-facing bud. Some exceptions are wall-trained and pillar plants where it is intended to induce growth in a more or less two-dimensional manner. Also, certain fruit trees, like the plum cultivar 'Victoria', have a naturally spreading habit and unless this is countered, branches will be weighed down and broken by the fruit crop. The amount of growth to remove (in other words, the length of stem which should be cut off), varies enormously but is almost entirely governed by the rapidity (sometimes called the vigour) with which the plant grows. In a table over the page I have summarised the routine pruning recommended for a wide range of common garden flowering shrubs, but advice such as that given here often seems to gardeners to beg two questions. Why should a dormant bud burst into life simply because the length of shoot above it has been removed and why should the amount of growth to remove relate to the general vigour of the plant. In other words, what is behind the old adage that the harder you prune, the more will new growth arise or, to put it more colloquially, that growth follows the knife?

The answer to these conundrums lies with the hormones that I described on p. 27 and with the phenomenon of apical dominance. The bud at the apex of a shoot produces an inhibitory hormone that diffuses back down the stem, permitting elongation of the shoot apex but suppressing the growth of lateral or side buds. Remove the distal or furthest part of the shoot and the source of the inhibitor disappears too. And, in general, the greater the length of main shoot removed, the greater will be the diminution of the apical dominance/lateral inhibition effect and, hence, apparently the more vigorous the response. Conversely, the removal of lateral shoots or buds increases the vigour of the shoot tip (and presumably its auxin content) and so limits future lateral bud break. Nonetheless, it is sometimes desirable to train a plant into a form that necessitates the lateral buds beginning to grow without removing the shoot tip. Perhaps the best

example is with a climbing rose grown against a wall. Removal of the main shoot will certainly stimulate lateral buds to grow, but all growth will be vertically upwards and the lower parts of the wall consequently remain bare of flowers. But by bending the main shoot downwards to a horizontal plane, the flow of inhibiting hormone down from the tip is curtailed almost as effectively as if the shoot had been severed.

Sometimes the objective of removing the apical part of a shoot by pruning is solely to encourage more leafy shoots from the lateral buds – hedges and hedge trimming are the obvious examples of this. But in most instances, pruning is performed to aid flower and fruit formation at the expense of leafy shoot development, or perhaps to influence the shape of the plant in a way that will be aesthetically pleasing but at the same time not diminish its flowering or fruiting. As flowers and fruits are usually borne on leafy shoots at certain periods of the year only, the timing of pruning therefore assumes critical importance. Careful analysis of the pruning regimes that I have given in the table will reveal a fairly general relationship between pruning time and flowering time. Most

By bending a branch to the horizontal or slightly lower, a climbing plant can be induced to produce flowers lower down and so ensure that the entire wall, and not just the top, is clothed with blooms.

Routine pruning of a range of common garden ornamental shrubs

Shrub	Pruning system	Shrub	Pruning system
Abutilon	Shorten the long shoots of *A. vitifolium* after flowering to maintain shape; cut back frost-damaged shoots in spring	Clematis (early species)	Cut back lightly after flowering, removing only weak growths or those outgrowing the allotted space
Berberis	Thin out old wood in the spring after flowering – old plants will stand hard cutting back; clip hedges after flowering	Clematis (mid-season)	Cut back fairly hard in February to strong new buds, consistent with keeping the plant within its allotted space
Buddleia	Cut back *B. davidii* to four buds from the base in early spring. Shorten shoots of other buddleias after flowering to improve shape	Clematis (late)	Cut back hard in February to strong new buds about 20 centimetres from base or from base of previous season's shoots, depending on amount to remove
Buxus	Normally needs no treatment when grown as a free-standing bush. Clip hedges in late summer and topiary also in early May	Cotoneaster	No regular treatment needed but rejuvenate old bushes by hard pruning in spring; February for deciduous types, April for evergreen ones
Calluna	Clip back old flowering shoots in early spring. Vigorous types will stand hard cutting back but replace them when very woody	Cytisus	Prune each year after flowering to shape the plants, but do not cut beyond the previous year's wood as the older wood will not produce new growth
Camellia	No regular treatment should be necessary, but old, misshapen and leggy plants can be cut back hard after the flowers fade	Elaeagnus	No regular treatment is needed, but thin out and shape plants in spring; February for deciduous types, April for evergreen ones
Ceanothus (spring-flowering)	As a bush, thin only in spring. Against a wall, cut back to within two buds of previous season's growth after flowering in spring	Erica	Remove the dead flower heads with shears; in April for summer, autumn and winter-flowering types, in June for spring-flowering forms
Ceanothus (summer-flowering)	Cut back hard to within two to six buds (less for the weaker shoots) in early spring – preferably in February unless very frosty	Escallonia	No regular treatment is needed for bushes, but wall-trained plants benefit from a shortening of lateral shoots in April
Chaenomeles	Cut out old wood, thin out and shorten side shoots after flowering and remove those inward and outward facing on wall-trained plants	Euonymus	Shape free-standing bushes during summer, clip hedges twice a year, trim back the upright shoots from prostrate forms in spring
Choisya	Cut out old wood and shape plants in spring; wall-trained plants may be cut back fairly hard after flowering	Forsythia	Against walls, cut out old flowering shoots after blossom has faded. In the open, thin out overcrowded shoots every three years
Clematis (early hybrids)	Cut back lightly in February to strong new buds, removing dead wood and other shoots to keep plant within its allotted space		

Shrub	Pruning system
Fuchsia	Large, hardy bushes are best given no regular treatment. Smaller and less hardy plants should have dead growth removed in spring
Hebe (Veronica)	No regular treatment needed but old bushes and those suffering frost dieback can be rejuvenated by pruning them hard in April
Hydrangea (mop-head)	Cut out the dead flower heads with three attached leaves after flowering in very mild areas but leave until the spring in most regions
Hypericum	Thin out the old wood and cut back previous year's growths to four buds in early spring. H. calycinum should be clipped with shears
Jasminum	Cut back flowering shoots of winter jasmine to within two buds of the base after flowering. Thin out summer jasmine after flowering
Kerria	Cut out the old wood immediately after flowering, pruning either to the base or down as far as vigorous lateral shoots
Laurus (bay)	Prune in spring, cutting back leggy bushes hard but vigorous young ones less so. Use secateurs if possible to avoid unsightly leaf damage
Lavandula	Cut back the old flowering shoots in late summer but renew old woody bushes as the plant does not respond well to very hard pruning
Magnolia	No regular treatment is needed. Shape large or distorted trees as necessary after flowering or in early autumn by removing entire branches
Mahonia	No regular treatment of any types of mahonia is needed but long, non-flowering shoots and old wood may be cut back in spring
Olearia	Prune O. haastii lightly to remove dead flowers in late summer. Old bushes can be rejuvenated by cutting back in April

Shrub	Pruning system
Philadelphus	Remove old flowering shoots from hybrid types and thin out generally after flowering. Rejuvenate old plants by pruning hard in spring
Potentilla	No regular treatment needed, but cut out old wood in early autumn after flowering. Old, overgrown shrubs will tolerate hard cutting back
Pyracantha	No regular treatment needed for bushes in the open. Against walls, shorten lateral shoots in summer and spring and train along wall
Rhododendron	Very carefully remove dead heads after flowering. Rejuvenate old bushes by hard pruning in spring and remove suckers from R. ponticum rootstocks
Ribes	No regular treatment needed, but old bushes of most flowering currants can be rejuvenated by hard pruning after flowering in late spring
Spiraea	Cut out old flowering shoots after flowering (spring-flowering types) and prune hard in early spring (late-summer-flowering types)
Syringa	Remove dead flower heads after flowering. Remove suckers annually. Rejuvenate old bushes by pruning back growth by up to a half in spring
Tamarix	Trim dwarf bushes or hedges of late-summer-flowering forms after flowering. Large old bushes may be pruned hard in early spring
Viburnum	No regular treatment needed although old wood may be cut out after flowering. V. tinus responds to hard pruning in spring
Weigela	No regular treatment needed, but remove old and leggy shoots and flowering shoots bearing no new growth after flowering
Wisteria	Shorten young whippy shoots back to six buds in July and then back to two buds from the base in December to encourage spur development

woody plants can be divided into two principal groups – those that bear their flowers and fruits on shoots produced during the previous year and those that bear them on the current year's growths. Most of the former flower in the early part of the season, generally before the end of June, but most of the latter do not produce enough growth sufficiently quickly to enable flowering to occur until after the end of June. A moment's thought will indicate that if the early-season flowering plants are pruned during the autumn, winter or early spring, the shoots bearing the flower buds will be removed. Hence they must be pruned in the late spring or early summer. But if the later flowering plants are pruned at this time, their flower buds will be removed. They, too, must be pruned, *after* they have flowered – either in the autumn, winter or early spring, before growth begins anew. There is much to be said for delaying the pruning of most of this group until early spring, for if the old shoots are left on during the winter they will help provide frost protection for the plant as a whole.

Whilst most of what I have said relates to perennial woody plants, it should not be forgotten that some forms of pruning can and should also be applied to annuals. Among ornamentals, routine dead-heading (always cutting back, of course, to a bud whenever possible) fulfils the criteria of removing dead and potentially diseased tissues, encouraging new flowering shoot development and, to some extent, of enhancing the shape of the plant too. Among annual edible crops, little if any pruning can or need be performed with the true vegetables, but vegetable fruits such as tomatoes, peppers, aubergines and the cucumber family all benefit from removal of some shoots for restraining overall growth or encouraging the plant's nutrient resources to be channelled towards the production of few large, rather than many small fruit.

Allied, at least to the training of plants, is staking, the artificial support of plant stems too weak to support their weight naturally. Among herbaceous garden plants, the need for staking usually arises because a particular cultivar has been selectively

Every young tree requires the support of a carefully positioned stake and purpose-made ties. Some trees, especially fruit trees on dwarfing rootstocks, need this support throughout their lives.

bred for flowers so large that the stem is too weak to hold them upright. Some of the giant-flowered show dahlias illustrate this particularly graphically. Among woody plants, there are generally two distinct reasons for the need to stake. The first is that a plant that occurs naturally as a multi-stemmed shrub has been pruned to form a standard with a large head on a single weak trunk; the June-flowering *Buddleia alternifolia* is a good instance of this. But the second and commoner reason arises when a tree that grows naturally among the protection of other trees in a woodland or forest is placed alone as a specimen plant in an exposed garden location. There it will be subject to the buffeting action of the wind and, although the structure of the stem is well equipped to tolerate at least some wind action, damge can ensue during winter gales. Evergreen plants that present a much larger surface area to the wind are apparently more prone to being damaged in this way although, in fact, their tissues are often more pliable. There are two quite distinct theories of staking; the first being

The honey scented, June-flowering Buddleia alternifolia is a classic example of a shrub trained on a single stem, which is more valuable and more attractive than many a small tree.

that a stake should be at least as tall as the tree itself and should remain in place for many years (and, in some instances, permanently). The second theory advocates the use of stakes much shorter than the tree and for a short initial period only. This is believed to encourage the development of resilient tissues in response to the bending action in the wind that the shorter stake allows. But whichever staking method is chosen, it is important that the stake is placed on the side of the tree facing the prevailing wind, so that the plant is blown away from the stake, not onto it where it will be subject to abrasion. And the tie securing the plant to the stake must be of a broad, expandable belt pattern, in order not to cut into the stem tissues.

Throughout this century knowledge has accrued constantly from scientific studies of the factors controlling the many biochemical processes that constitute plant growth. Some of these findings have found important practical applications in horticulture and, whilst most are perhaps of relevance only to commercial growers, there are some significant benefits from these findings to be derived by gardeners too.

I made reference in Chapter Two to growth-regulating chemicals or hormones, those substances that stimulate or retard cell growth or division and so govern the rate at which plants grow and, more especially, dictate when and where new roots, stems, flowers and other tissues form. The most important practical gardening application of plant hormones or synthetic analogues of them is in the aiding of root development when striking cuttings (p. 129). But two other types of hormone-like chemicals are available to gardeners and have aroused considerable interest in recent years.

The first is represented by the substance 'dikegu-lac' which is formulated as a liquid concentrate for regulating the growth of hedges. Dikegulac is now the only growth retardant available to gardeners and it functions by being absorbed through the foliage, retarding the growth of the leading shoots and thus reducing apical dominance in a way comparable with the removal of the shoot tip by

There are no viable alternatives to regular trimming if a hedge is to be kept neat and attractive. Growth retardants have some value but their use is limited to certain species.

pruning (p. 103). Care is needed with the timing and rate of application of dikegulac on different hedge species and the material must not be used on hedges younger than three years, nor on certain types of plant (most notably yew and box among common hedging species).

Commercial horticulturists employ a range of other growth retardants for various purposes. And it is because such dwarfing agents as daminocide or paclobutrazol are unavailable to gardeners that the amateur growing of dwarf pot chrysanthemums, for instance, is impossible. The short-lived effect of such growth retardants is well demonstrated however if such dwarf plants are grown on for a second season when they form tall, rangy and uselessly unattractive objects.

The second type of hormone that has found limited use in the garden is the fruit setting agent 2-naphthyloxy acetic acid which is available as an aerosol for spraying the flowers to improve productivity in tomatoes or other crops. Strange and sometimes bizarre effects of fruit malformation can arise through over-dosing with setting agent and it is unlikely to be necessary in the average garden greenhouse if general aspects of crop husbandry are attended to carefully.

My mention of the hormonal regulation of growth in dwarf chrysanthemums brings me on to another aspect of horticultural physiology, for the florist's chrysanthemum is the classic example of a plant whose flowering rhythm is manipulated artificially. As I explained on p. 28, chrysanthemums are short-day plants and need twelve hours of daylight or less in order for flower initiation to commence. Grown outdoors in a northern hemisphere mid-temperate garden, they receive this short-day stimulus at a time of year that leads to flowers opening during the autumn. Their days can be artificially elongated, however, by supplying them with additional illumination. This retards flower initation and enables the plants to be offered in bloom at Christmas (or indeed, by careful manipulation of light, at all other times of the year too). The discovery that it is in fact the long length of the night or dark period rather than the short length of the day that governs the process has had an important practical application. Instead of expensively providing continuous light and so lengthening the day, the dark period can be broken simply by exposing the plant to short periods of illumination during the night. Because of the impossibility of inducing dwarfing however, artificial control of flowering in chrysanthemums is of little direct interest to gardeners. Nonetheless, a somewhat comparable process in another popular pot plant can be attempted. The poinsettia, *Euphorbia pulcherrima*, only produces the attractive red colouration to its bracts when subjected at a certain stage of its growth to a dark period of fourteen hours. I have outlined on the right the procedure to be followed in order to obtain a second year's benefit from a poinsettia obtained as a commercially grown plant at Christmas.

Having heard something of the effect that abnormal lighting periods can have on plant growth, gardeners are sometimes concerned when their gardens are illuminated by street lighting for much of the night. Usually the intensity of the illumination is too low to have any appreciable effect, but sometimes retarding of flowering in short-day

Timetable for treatment of poinsettias obtained at Christmas to achieve a second year's flowering

January–February	As soon as the leaves drop, cut back stem to about 10 centimetres
February–May	Keep pot in cool, frost-free place and allow compost to become almost dry
May	Bring plant into warmer conditions and water. Remove from pot, carefully pull away some of the old compost from the periphery of the pot-ball and re-pot in one pot size larger, using fresh, peat-based compost
May–September	Water regularly and feed weekly with liquid tomato fertiliser; remove excess shoots to leave the four strongest
End September–end November	Cover the plant completely every evening with a black plastic bag; remove in the morning after fourteen hours of dark treatment
End November–Christmas	Continue to feed and water at room temperature and in good light but not direct sunlight

Varying degrees of automation are employed in commercial glasshouses so that poinsettias undergo the dark period necessary for red colouration to develop.

Compromises must be made in order to satisfy the desire for year-round availability of garden produce while still maintaining cost-effectiveness in producing it.

plants or enhancement of flowering in long-day plants can be attributed to this cause. Perhaps the commonest effect arises when a long-day flowering tree or shrub actually has some of its branches enveloping a street lamp which causes that part of the plant to burst attractively into unseasonal bloom.

I have mentioned elsewhere some of the art and science of ensuring that the productiveness of fruit and vegetables can be spread as uniformly as possible through the season – by careful choice of varieties and by careful planning of sowing times. However, much of the effort can be wasted if the maximum use is not made of the potential offered by modern methods of storage and corresponding care is not given to the harvesting and handling of the produce.

All-year-round availability of fruit and vegetables is today taken for granted. Two developments within the past fifty years or so have brought about this revolutionary change. Modern intercontinental transportation now enables supermarkets to offer such previously undreamed of (and, in reality, quite ludicrous) temptations as fresh Guatemalean mange-touts for Christmas. In consequence, some gardeners may have been prompted to grit their teeth, pay an inflated price and feel less inclined to

bother storing their own vegetables. This is unfortunate, partly because many garden crops can be stored in a fresh condition fairly easily, and also because the second modern development, the very widespread possession of a home freezer, enables some garden crops previously very seasonal in their availability to be eaten all year round.

It is beyond my brief to enter into the technology of home freezing, other than to urge the careful selection of soft fruit and vegetable cultivars, for some undoubtedly freeze much more satisfactorily than others. (Modern seed catalogues indicate the best cultivars for freezing.) Nor is it appropriate to describe in detail the more traditional procedures such as pickling, salting or bottling. I have therefore listed opposite my suggestions for the most useful methods to adopt with the commonest fruit and vegetable crops. However, some of the methods that I suggest for the storage of produce in a fresh or unaltered state do entail operations that in effect are extensions of gardening plant husbandry, and certain general principles are appropriately discussed here.

There is a major distinction to be made between frost-hardy and frost-tender produce and between bulky or weighty crops and smaller, tender, fragile ones. Hardy produce, especially bulky hardy produce such as most leafy brassicas, beetroot, carrots, leeks, parsnips, swedes and turnips should be stored where they grow, in the ground – remember that in the case of the root crops, the structure being stored is serving the role naturally intended for it (p. 16). Nonetheless, once temperatures fall below about −5 °C, some tissue damage may be expected with some cultivars and this will generally lead to rotting when the temperature rises again and fungal and bacterial invasion of the wounds occurs. However, in passing, I should add that the ability of carrots in particular to survive very much lower temperatures during the British winters of 1985–6 and 1986–7 surprised many gardeners, this one included. Nonetheless, some additional protection is desirable in colder areas and is best provided by a layer of straw about 15 centimetres thick over

the crowns, lightly covered with soil for anchorage. When protecting root crops in this way, it is sensible to delay laying the straw until mid-December, for the extra warmth that it provides can encourage the plants to continue growing to their ultimate detriment. It must also be said that a protective winter covering of this type can attract slugs, and in gardens where slug infestation is already bad a balance must be struck on the basis of personal experience between the losses to be expected from the slugs and the losses from frost.

Almost all other vegetables and fruit are not hardy and must be protected from frost if they are to be stored in their natural state, rather than preserved. But whilst very few will survive at freezing temperatures, nor will they survive at elevated ones, and the general principle of successful prolonged storage is to keep them cool and fairly dry. Low temperatures not only slow down or prevent tissue growth in active organs such as lettuce heads, but also minimise deleterious chemical changes in mature fruits such as the conversion of sugars back to starch that takes place in sweet corn cobs. Once harvested therefore, the more rapidly that produce can be cooled the better – it is unwise to cut or harvest all of a crop at once if your storage space can only accommodate half of it.

Bulky, non-hardy produce, such as onions, potatoes, apples and pears, are best stored in a dark, well-ventilated room or building with as uniform a temperature as possible, as the table indicates. Smaller and more-fragile produce is best kept in a refrigerator at 2–4 °C. However, some non-hardy crops are adversely affected by temperatures in excess of freezing but below about 10–12 °C, a phenomenon known as chilling injury. Warm-temperate or sub-tropical crops, such as cucumbers and other cucurbits, egg plants, peppers, tomatoes and many types of bean, have a fairly limited storage life therefore, and some indications of the periods over which they might reasonably be expected to stay fresh and edible are given on the right. It must be added, however, that these times refer to mature produce and it is often possible,

Best storage methods for common fruit and vegetables[a]

Soft fruit	
Black currant	
Gooseberry	
Loganberry and similar hybrids	
Raspberry	Bottled or frozen
Red currant	
Strawberry	
White currant	

Top fruit	
Apple	Fresh, at as uniform a temperature and as close as possible to 3–4.5 °C; either singly, unwrapped on slatted shelves and not touching, or in lightweight clear polythene bags, loosely folded with each containing about 3 kilograms of apples and with one puncture hole for each 0.5 kilogram of contents
Damson	Bottled or frozen after removing stones
Peach	Bottled or frozen after removing stones
Pear	Fresh, at as uniform a temperature and as close as possible to 1 °C; singly, unwrapped on slatted shelves, not touching; check regularly for ripeness
Plum	Bottled or frozen after removing stones

Vegetables	
Asparagus	For short period only in refrigerator
Bean (broad)	Frozen
Bean (French)	For short period only in refrigerator
Bean (runner)	Frozen or salted
Beetroot	In ground, where grown
Broccoli	Frozen
Brussels sprout	Frozen
Cabbage (spring)	For short period only in refrigerator
Cabbage (summer)	For short period only in refrigerator
Cabbage (winter and savoy)	For short period only in refrigerator
Calabrese	Frozen
Carrot	In ground, where grown
Cauliflower	Frozen
Celery (self-blanching)	For short period only in refrigerator
Celery (trenched)	In ground, where grown
Courgette	For short period only in refrigerator
Cucumber	For short period only in refrigerator
Kale (curly)	For short period only in refrigerator
Kohl-rabi	Frost-free in fairly dry conditions
Leek	In ground, where grown
Lettuce	For short period only in refrigerator
Onion (bulb)	Frost-free, in nets after drying
Onion (salad)	For short period only in refrigerator
Parsnip	In ground, where grown
Pea	Frozen
Potato (early)	Not worth storing
Potato (maincrop)	In paper sacks at as uniform a temperature and as close as possible to 10–15 °C
Radish	For short period only in refrigerator
Shallot	Frost-free, in nets after drying
Spinach	Frozen
Swede	In ground, where grown
Sweet corn	Frozen
Tomato	Unripe, on plants, hung upside down in cool place
Turnip	In ground, where grown

[a] it must be stressed that not all cultivars of any particular plant are suitable for all storage methods and that the methods suggested here are those most likely to retain the produce closest to its original condition and flavour. Of course, many fruit and vegetables *can* be stored in other ways as preserves, jams, chutneys and so forth

One of the most important rules of storing produce is to check it regularly before possible contamination occurs. Some hardy produce can be stored in the ground, where it grows.

with care, to keep immature green tomatoes, for instance, for much longer, removing them at intervals to ripen in warmth.

It is essential with any storage, as opposed to preserving system, to ensure that only sound produce is used. Any affected by pests or diseases will deteriorate rapidly and quite possibly spread the damage to previously healthy produce. Take especial care when selecting apples and pears for storage only to use those picked from the tree, with the stalk intact. Any windfalls or other bruised samples should be rejected. Once stored, inspect fruit or other accessibly arranged produce regularly and carefully remove any damaged specimens. Never use fungicide or insecticide treatments on harvested produce to attempt to prolong storage life, and try to harvest slightly immature fruit which will usually keep rather longer.

Period over which harvested vegetables can be expected to remain fresh when stored at 0–1 °C or in a refrigerator at approximately 4 °C

Vegetable	Maximum storage period (days)	
	In refrigerator at about 4 °C	*At 0–1 °C*
Asparagus	4	14
Beans (runner and French)	8	–
Broccoli	5	14
Brussels sprout	7	14
Cabbage (spring greens)	7	7
Cabbage (winter white)	60	120
Calabrese	3	3
Cauliflower	4	20
Celery	15	60
Courgette	14	–
Cucumber	14	–
Kale	7	15
Leek	15	60
Lettuce	7	14
Marrow	60	–
Onion (salad)	5	14
Pea (unshelled)	14	14
Pepper (*Capsicum*)	21	–
Pumpkin	60	–
Spinach	3	14
Tomato	14	–
Watercress	1	4

CHAPTER SEVEN

Plant variety and plant reproduction

'Variety is the spice of life': a well-stocked border not only gives visual pleasure but also points to the sheer diversity of plants available to the modern gardener.

My memory does not serve me well enough to recall exactly what William Cowper had in mind when he offered the observation that 'Variety's the very spice of life that gives it all its flavour', but as the same gentleman also provided the view that 'Who loves a garden loves a greenhouse too', I find it endearing to imagine that he had had some horticultural inspiration. For surely it is the very variety of plant life that elevates a garden above the humdrum and that provides so many visual and other pleasures. As long as men have been aware that living things do differ one from the other, and especially since they first needed to refer to and distinguish the edible from the inedible or poisonous berries and leaves that they collected, the naming of organisms has had a valid *raison d'être*.

However, some 10 000 years ago, our ancestors began to cultivate rather than merely to hunt and gather, and subsequently they came to be aware of and later still to manipulate articifically the phenomenon of plant reproduction. Then, not only did the ability to distinguish and name become important, but some grasp of the relative relatedness of organisms assumed significance too. Today, the whole subject is embodied in the study of classification, in the science of taxonomy which even now serves, or attempts to serve, two quite separate purposes. The first is to give every living thing an unambiguous name, immediately understandable to anyone, anywhere in the world. The second is to group together organisms that are closely related. (There is, in practice, a third purpose in that most taxonomists hope that their naming and classification systems give an indication not just of modern relationships but of evolutionary progressions too.)

The naming system devised by the Swedish botanist Carl von Linné (usually known as Linnaeus, 1707–78) is now used universally. It has the merit of being international through being based upon one language, Latin (although names are sometimes borrowed from Greek or modern languages and Latinised), and also of assisting the grouping together of closely related types by giving each organism a dual name or binomial. I shall shortly give some examples of these binomials. However, to appreciate fully their usefulness we must first delve a little deeper, for the basis of Linnaeus' system is the species, at once one of the most valuable and important concepts in all of biology but also one of the hardest to define and appreciate. My preferred definition of a species is 'a group of individual organisms that are readily able to breed among themselves but not generally able to breed with organisms belonging to other groups or species'. A species, therefore, is reproductively isolated. A group of species that displays similarity in many ways is called a genus (plural genera), and in the instances when members of one species *do* prove able to breed with those of another, it is almost always with those species closely enough related to be placed in the same genus.

The binomial of each organism comprises a genus name and a specific or trivial name. For example, the plant known to the colloquial English speaker (but not to anyone else) as the dog rose is given the binomial *Rose canina*. This unique combination of the generic name *Rosa* with the specific name *canina* conveys immediate meaning to any botanist, anywhere. It also indicates that this plant is very similar (closely related if you wish) to such other plants as *Rosa lutea*, *Rosa virginiana* or *Rosa watsoniana*. (For brevity, the generic name is often abbreviated to the initial letter if the genus is repeated – e.g. *R. lutea,* and if the genus but not species is known the abbreviations sp. or spp. for singular or plural species are used.) Often, the specific name attempts to give some brief diagnostic information about the plant – *lutea* suggests a rose with yellow flowers or *virginiana* one originating in Virginia, for example – although, as in *watsoniana*, it may do no more than commemorate someone like Watson who might have played some part in its study. Sometimes, as in *canina*, the name is quite fanciful.

Genera that appear similar or closely related are grouped in turn into families, and gardeners will be

The Systema Naturae devised by the eighteenth-century botanist, Linnaeus, for the naming and classification of plants and other organisms is still in universal use today.

familiar with many important examples – the rose family Rosaceae, the pea and bean family Leguminosae, the huge daisy family Compositae, the grass family Gramineae and, largest of all with over 20 000 species, the orchid family Orchidaceae are but a few of them.

The only taxonomic groups larger than the family that are likely to be of interest to gardeners are the two major divisions of flowering plants called monocotyledons and dicotyledons, to which I have referred already (p. 12). However, there are other taxonomic categories of considerable importance to anyone whose interests are more horticultural than botanical, for relatively few pure species are grown routinely in gardens. Below the species in classificatory rank lies the subspecies (abbreviated to ssp.), a group of individuals that has developed slight (and usually morphological or behavioural) differences by evolutionary change, generally brought about by geographic isolation from the main species. Commonly, subspecies arise on small offshore islands, but they remain fertile with the main species on the nearest mainland should the opportunity for inter-breeding arise. If the opportunity is denied and isolation persists for a long time, this inter-fertility may be lost and the subspecies then becomes a species in its own right. A group of individuals within either a species or a subspecies that differs from the remainder of the group in smaller or less biologically significant ways is called a variety (var.). Many garden plants are varieties that have arisen naturally and been collected and propagated for some desirable horticultural feature – the cauliflower, cabbage and Brussels sprout are all varieties of the same species *Brassica oleracea*, for instance. Even more significant, however, is the cultivar (cv.) – the variety that has arisen, not naturally, but in cultivation, either by accident or by deliberate artificial hybridisation and selection. Cultivars can exist within species (*Calluna vulgaris* 'H. E. Beale'), subspecies (*Juniperus communis* ssp. *depressa* 'Echiniformis') or even within varieties (*Brassica oleracea* var. *capitata* 'Golden Acre').

The wild cabbage in its natural habitat on cliff-tops bears little resemblance to the many and varied forms of the same species that are familiar in cultivation.

Sometimes, however, the logic of Linnaeus and his modern disciples seems obscure. A glance at the vegetable garden will indicate several plants superficially similar – cabbages, lettuces and spinach for instance. Yet all three are placed in quite distinct families and each shares its family with some improbable companions – the cabbage with the nasturtium, the lettuce with the sunflower (and many tropical trees) and spinach with a bizarre mixture that includes some desert shrubs and climbers. The reason for this apparently illogical state of affairs is because Linnaeus chose not the leaves or other vegetative parts of plants as a basis for classification, but the structure of their flowers. He did this largely because the flowers are much less prone to vary under environmental influences (no matter how good or poor the growing conditions, the flowers of any species are usually constant in size and shape). The flowers, moreover, are the reproductive organs of a plant, through which and by which the overall range of variety in the plant kingdom arises. And because I hope that gardeners will wish not only to appreciate the way that this happens but will also want to undertake seed collection and possibly some controlled plant breeding too, a little explanation of flower function is pertinent.

On p. 22 I described in outline the constituent parts of a flower, including the stamens which have the function to produce pollen. The pollen contains male cells and it must somehow be transferred to the stigma that contains the female cells. This process is called pollination and is the inevitable prelude to fertilisation – the actual fusion of male and female cells which is followed, subsequently, by the development of seeds. Sometimes, when stamens and stigma are produced on the same flower, there is more or less physical contact between them and the pollination process is very simple, although visits to the flower by bees or other insects seeking nectar can help its efficient functioning. When stamens and stigma are separated on different flowers or different plants (or even when male and female cells produced on the

same flower are incompatible and unable to fuse), some external transfer process is needed. Insects are the most familiar carriers of pollen, and the structure and colouration of many flowers is ingeniously contrived to attract insects and facilitate their pollen gathering and transfer. In temperate climates the wind is the other great carrier of pollen, but in tropical plants pollen may be carried by a wide range of other means, including hummingbirds, bats and slugs. The structure of a wind-pollinated flower is quite different from an insect-pollinated one, and the amount of pollen produced is very much greater in wind-pollinated flowers, for the transfer is much more of a hit and miss process.

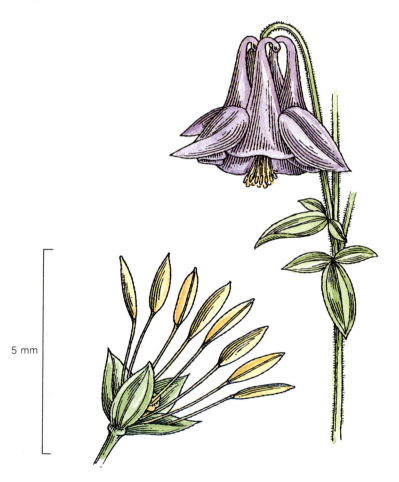

The contrast between a wind-pollinated flower (left) and an insect-pollinated one (right). The wind pollination process requires that the stamens are exposed and also that a great deal of pollen is produced. Wind-pollinated flowers are also usually dull coloured, having no need to attract insects.

5 mm

In passing, I have referred already to those plants on which the male and female cells are incompatible, where pollen from one plant must be transferred to a stigma on another. Such cross-pollinating plants are called outbreeding, in contrast to those which manage very well with their own pollen and are called self-pollinating or inbreeding. The ability or otherwise of a plant to be cross- or self-pollinated is of more than academic interest because gardening today is rather less about species as they occur naturally than about hybrids, generally produced artificially.

Before discussing the nature of hybrids in rather more detail, I must mention one group of cross-pollinating garden plants for which the effectiveness of pollination is especially important. Most fruit trees are more or less self-incompatible and require pollen from another tree of a different cultivar if they are to bear fruit. But among apples in particular, the situation is complicated because some cultivars are much better pollinators than others and not all cultivars are in flower at the same time. Moreover, one popular cultivar, 'Bramley's Seedling', is a genetic triploid (containing three times the basic number of chromosomes for the organism) and requires not one, but two additional trees in order for successful pollination to occur. In some fruits, most notably cherries and pears, there are distinct groups within which the cultivars are compatible but between which they are not. I have listed overleaf some suitable combinations of fruit tree cultivars to obtain satisfactory fruit set.

The sole purpose behind the diverse colours, forms and perfumes of plants is to attract insects (bees, most importantly) in order to effect pollination.

A few types of plant such as the apple 'Bramley's Seedling', are genetically triploid. Three individual plants of appropriate variety are needed before fertilisation can take place.

Suggested combinations of some popular fruit tree cultivars to obtain good pollination

Fruit	Suggested pollinator	Fruit	Suggested pollinator
Dessert apples		'Doyenné du Comice'	'Conference' or 'Onward'
'Beauty of Bath'	'Greensleeves' or 'Idared'	'Louise Bonne de Jersey'	'Conference'
'Blenheim Orange'	'Discovery' or 'Greensleeves'	'Onward'	'Conference' or 'Williams' Bon Chrétien'
'Cox's Orange Pippin'	'Discovery', 'Greensleeves' or 'James Grieve'	'Williams' Bon Chrétien'	'Conference' or 'Onward'
'Crispin'	'Cox's Orange Pippin' or 'Discovery'		
'Discovery'	'Cox's Orange Pippin' or 'Greensleeves'	*Plums*	
'Ellison's Orange'	'Cox's Orange Pippin' or 'Greensleeves'	'Cambridge Gage'	'Czar' or 'Marjorie's Seedling'
'Fortune'	'Discovery', 'Greensleeves' or 'James Grieve'	'Czar'	Self-fertile
		'Marjorie's Seedling'	Self-fertile
'Greensleeves'	'Discovery' or 'Grenadier'	'Rivers Early Prolific'	'Czar' or 'Victoria'
'Idared'	'Cox's Orange Pippin' or 'Discovery'	'Victoria'	Self-fertile
'James Grieve'	'Cox's Orange Pippin' or 'Discovery'		
'Jupiter'	'Discovery', 'Spartan' or 'Sunset'	*Damsons*	Self-fertile
'Kent'	'Cox's Orange Pippin' or 'James Grieve'		
'Laxton's Superb'	'Cox's Orange Pippin' or 'Greensleeves'	*Cherries*	
'Redsleeves'	'Fortune' or 'Greensleeves'	'Morello' (cooking)	Self-fertile
'Spartan'	'Discovery' or 'Greensleeves'	'Stella' (sweet)	Self-fertile
'Sunset'	'Cox's Orange Pippin' or 'James Grieve'	*Apricots*	
'Tydeman's Late Orange'	'Greensleeves' or 'Spartan'	'Moorpark'	Self-fertile but use a paint brush to aid set
'Worcester Pearmain'	'Cox's Orange Pippin' or 'Greensleeves'	*Nectarines*	
Cooking apples		'Lord Napier'	Self-fertile but use a paint brush to aid set
'Bramley's Seedling'	'Discovery' plus 'Spartan'		
'Grenadier'	'Discovery' or 'Greensleeves'	*Peaches*	
'Howgate Wonder'	'Ellison's Orange' or 'Tydeman's Late Orange'	'Peregrine'	Self-fertile but use a paint brush to aid set
'Lord Derby'	'Spartan' or 'Tydeman's Late Orange'		
'Rev. W. Wilks'	'Idared'	*Medlars*	
Pears		'Nottingham'	Self-fertile
'Conference'	'Louise Bonne de Jersey' or 'Onward' ('Conference' will set some fruit on its own but is always better with a pollinator)	*Quinces*	
		'Vranja'	Self-fertile

A hybrid is the offspring of two genetically unlike individuals. These individuals may be cultivars, varieties, subspecies, species or, very occasionally, genera. Often they are sterile and thus unable to reproduce sexually (the horticulturist must therefore perpetuate them by vegetative means (p. 127)). Why are hybrids so important to a gardener? Hybridisation, the production of hybrids by cross-fertilisation, offers a means of combining the attributes of two separate plants in one individual. Perhaps there are two fairly similar (and genetically closely related) species, one having large white flowers and the other smaller red ones. Cross the two artificially, collect the seed, grow the progeny and select seed from those individuals that have the largest and reddest flowers, repeat the process several times and in due course you will have a population of plants consistently having flowers both large and red. My example is simplistic but serves, I hope, to demonstrate the principle.

Hybridisation was once referred to as 'improving', and the use of this term reveals much about human nature for we are never satisfied with what nature has supplied but seek constantly to change it to meet our own criteria of desirability and excellence. Generally, a hybrid, be it a plant or a mongrel dog, tends to be larger and in some way more vigorous than its parents. But a hybrid plant allowed to perpetuate itself by further cross-pollination among its own individuals will still be fairly variable. In recent years, even this variability has been found a disadvantage, especially by the commercial growing fraternity who, for reasons of marketing and ease of harvesting, consider it desirable that crops should not only be large, but also uniformly large and maturing at a uniform time. Such demands have led to the development of commercially viable F_1 hybrid cultivars, and this designation now appears increasingly in seed catalogues, usually with some appended implication that they are invariably better than the alternatives. The way in which an F_1 (or first filial generation) hybrid differs from an open-pollinated plant is outlined below.

A comparison between the breeding programmes adopted to produce F_1 hybrid and open-pollinated cultivars. Hybrid production is very time-consuming and expensive and not possible or appropriate for all types of plant.

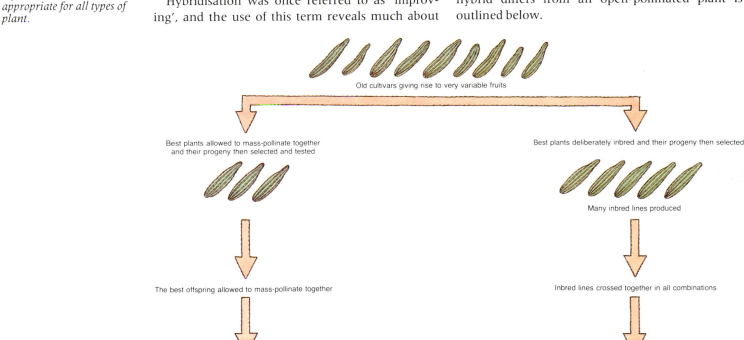

Old cultivars giving rise to very variable fruits

Best plants allowed to mass-pollinate together and their progeny then selected and tested

Best plants deliberately inbred and their progeny then selected

Many inbred lines produced

The best offspring allowed to mass-pollinate together

Inbred lines crossed together in all combinations

After many selections, a new open-pollinated cultivar

Progeny tested, and the best of them give new hybrid cultivars

Despite what catalogues may imply, the F_1 hybrid is not necessarily always better than the open-pollinated one and there are several features of such hybrids that are worth considering. The uniformity so beloved of the commercial grower who wishes to harvest an entire field of cabbages with one pass of his automated cutting machine may not be so desirable for the gardener who prefers his cabbages to mature more irregularly over several weeks. And the giant flower heads of such bedding plants as some of the F_1 hybrid African and French marigolds are not to everyone's liking, splendid as they may appear and appropriate as they may be on a municipal traffic island. Size and uniformity should be considered carefully, therefore, before you place all of your eggs in the basket of the F_1 hybrid.

Almost invariably, F_1 hybrid seed is more expensive than that of open-pollinated cultivars because its production is costly and labour-intensive. The inbred lines must be maintained carefully, away from any possible contaminating pollen. And the actual crosses between the inbred lines must usu-ally be performed very carefully by trained hand to avoid the transfer of pollen onto a stigma of the same plant. Even then, such tricks as bud pollina-tion (transferring the pollen to the stigma of an unopened flower before its own pollen-bearing stamens have matured) or pollination of emascu-lated flowers from which the stamens have been removed may be necessary. An alternative to hand-pollination is to use insects but breed genes for self-incompatibility into the inbred lines, although this itself is a lengthy and expensive process. Plants that naturally are markedly self-pollinating, such as lettuces and French beans, are extremely difficult to develop as F_1 hybrids, and the results would almost certainly not repay the effort involved. In yet other instances, such as radishes, there is already accepta-ble uniformity in the open-pollinated cultivars.

Before leaving the subject of F_1 hybrids, how-ever, two additional points should be made. Shortly, I shall be describing the excitement of saving seeds and also of making crossings and selections from your own garden plants. It must be said, however, that saving seed from an F_1 hybrid is a waste of time for it will almost certainly result in a total hotch-potch of very variable plants, the result of cross-pollination within the population and with other cultivars growing nearby. But if the very high degree of uniformity and very high cost of F_1 hybrids does not appeal to you, the greater varia-bility and lower cost of the F_2 hybrid might, espe-cially with such plants as pelargoniums for which the seed of open-pollinated cultivars is unavailable. And F_2 hybrid is produced by carefully allowing a population of F_1 hybrid plants to pollinate together freely, but still preventing any cross-pollination with other cultivars.

The end results of flowering and pollination are seeds. I happen to think that a packet of seeds still represents almost the best value in gardening but, annoyingly, seed companies often include more seeds of some types of plant in the packets than even the largest garden could cope with. In most cases it is possible to save seeds from one year to the next, and I have indicated in the table opposite the

The merits of F_1 hybrids are most apparent in commercial production where it is desirable for all individuals in a crop to mature simultaneously. This may be less desirable in gardens, however.

storage time over which some of the commoner vegetable seeds may be expected to retain their viability once the sealed metal foil packet has been opened. However, gardeners who buy vegetable seeds anew each season do obtain the benefit of legal protection which guarantees minimum germination percentages for seeds. These percentages are listed below.

Buying seeds in a packet, sowing what could pass for brown dust or grains of sand, and watching them grow ultimately into plants to grace your garden is certainly exciting for any lover of plants. Actually to collect the seeds from plants that you have grown is even more satisfying, but to collect and sow seeds that have formed following a deliberate and careful cross-pollination that you have

The F_1 hybrid plant has advantages – in pelargoniums for instance it means they can be raised readily from seed. However, the older vegetatively propagated varieties like this 'Irene' then tend to fall from favour.

Period over which some common vegetable[a] seeds may be expected to retain their viability once the seed packet is opened[b]

Seed	Storage period (years)
Beans (all)	1
Cabbage	10
Carrot	6
Cauliflower	6
Leek	3
Lettuce	3
Onion	3
Parsnip	3
Radish	10
Swede	10
Tomato	10
Turnip	9

[a] very few controlled studies have been made of ornamental seeds
[b] at room temperature (about 20 °C) and humidity. For every 5 °C drop in temperature down to 0 °C, and for every 1 per cent decrease in seed moisture content from about 15 per cent down to about 5 per cent, the storage period is doubled. The times quoted can be greatly extended, therefore, by storing seeds in screw-top jars with a sachet of moisture-absorbing silica gel in a refrigerator

Minimum legally guaranteed percentage germination of vegetable seeds[a]

Seed	Guaranteed percentage germination
Bean (broad)	80
Bean (French)	75
Bean (runner)	80
Beetroot	70[b]
Brussels sprout	75
Cabbage	75
Calabrese	70
Carrot	65
Cauliflower	70
Celery	70
Cucumber	80
Leek	65
Lettuce	75
Marrow	75
Onion	70
Parsnip	no legal minimum
Pea	80
Radish	70
Spinach	75
Tomato	75
Turnip	80

[a] no legal guarantee is available for ornamental seeds, and it should be appreciated that whilst these values give useful indications of the relative results you may expect in your garden, they are based on the germination to be obtained under defined and controlled test conditions
[b] except 'Cheltenham Green Top' (50%)

performed is the most appealing of all. Collecting seeds from plants that have grown and flowered ad lib in the open garden can be a chancy but adventurous exercise. Self-pollinating species like peas or cross-pollinating plants grown in isolation from related types may be expected to give progeny reasonably similar to their parents – what is generally known as 'true to type' or 'coming true from seed'. Nonetheless, isolation is a relative term, for the wind can blow pollen many kilometres and bees may wander up to several kilometres on their foraging flights so there can be no certainty that alien pollen has not brought about a particular fertilisation. The young plants may differ therefore in some significant (and just conceivably more appealing) way. Provided the seed heads are collected just before maturity when the seed is discharged or spilled, they can then be allowed to complete their ripening. Most dry seeds are best allowed to complete their ripening in paper bags kept in a dry, well-ventilated place and then stored as I shall describe. Seeds like tomatoes and marrows, for instance, borne within soft, fleshy fruits should first be washed free from the sur-

rounding tissues before being allowed to dry at room temperature. Careful washing away of the fleshy tissue is important, for sometimes this contains inhibitors intended to prevent the seeds from germinating while still attached to the parent plant.

Although it is perfectly possible to make artificial crosses between plants growing in the open garden, the more carefully controlled conditions of the greenhouse make this a better environment for those plants small enough to be grown in pots and transported indoors. I have indicated diagrammatically below the basic techniques of pollination and of ensuring that there is no likelihood of contaminant pollen being introduced. The procedures will vary slightly from plant to plant because of differences in flower structure and differences in the maturing times of male and female structures. And although I do not have the space to describe the procedure for a complete breeding programme in detail, it may be said that, in general, it is wise to concentrate on one particular feature in the offspring raised from your first crosses and attempt to improve that by further crosses. Perhaps some flowers for instance, are rather redder than those of

Plant breeding is a precision task. The tools for hand pollination are the forceps, the soft brush, scissors and a beaker of meths for sterilising the equipment between pollen transfers. Perhaps most important of all are a steady hand and a good eye to remove the pollen-bearing stamens from the flower of one plant, such as this lily, to transfer them to the flower of another.

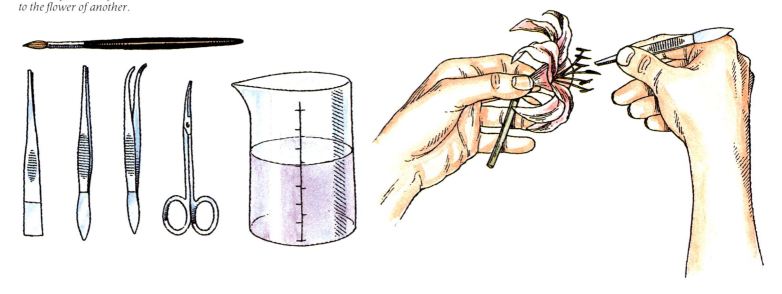

their parents or their siblings. Try therefore crossing together these red types rather than trying to take account of the fact that some also have larger leaves. Of course, the more that you become involved with the whole exercise, the more you will be absorbed by it and will want (and need) to learn some of the underlying genetic principles. Do not give up too soon if your results are frustrating, for desirable characters may be inherited in what is termed a recessive manner – they may be present in only a few individuals and may not always occur in every successive generation. Characters may be linked genetically one to another – it may prove impossible, for instance, to produce plants that have both blue flowers and strong scent. But above all, be assured that professional plant breeders are faced with just the same problems and will encounter just the same excitements and disappointments – and if you choose to work with a genus of fairly neglected or out of the ordinary plants, you may well find yourself soon becoming an authority on the subject and making a really worthwhile contribution. Perhaps a discussion with one of the major seed companies before you start serious crossings and selection would point you in the direction of a neglected group of plants with plenty of natural variability and, most important, reliable seed production and a fast generation time. Trying to improve the quality of Scots pines, for instance, could prove a frustrating discipline.

Seeds, even good fresh seeds, will not necessarily all produce plants. Sometimes a collection of seeds includes some that are immature or non-viable because of a partial failure of germination, but these are usually fairly easily recognisable by their small size and generally shrivelled appearance. It is unlikely that you will encounter many in commercial packeted seed for they are removed by grading machines. Nonetheless, some seeds do consistently offer low germination percentages and this is most commonly because their embryos require a further period of development which they are denied in the efficiency of modern harvesting and marketing and the rapidity with which sowing follows both.

Even when these constraints have been taken into account, the needs of individual types of seed must be considered carefully, for each has its optimum conditions of temperature, light, air, moisture and other factors before it will germinate. Generally, the advice given on modern seed packets is both reliable and helpful, although gardeners can sometimes be left confused, for the instructions for sowing can vary from one seed company to another.

In general, my advice is to follow the guidelines given on seed packets or in seed catalogues. To repeat all that advice here would be pointless, but an explanation of and the reasoning behind some of the required conditions should be helpful, and I have listed below the conditions that may assist the germination of some of the more difficult seeds among common garden plants. Perhaps the greatest number of problems attach to the phenomenon of

Special conditions needed to assist the germination of some common and traditionally difficult garden seeds

Seed	Germination conditions
Alstroemeria	(1) Sow 65 millimetres deep (2) 7 days at 21 °C (3) 21 days in refrigerator (4) Return to 21 °C
Begonia	(1) Sow on surface of compost (2) Minimum temperature of 21 °C with good light
Canna	(1) Abrade seed coat carefully (2) Soak in water for 48 hours (3) Minimum temperature of 21 °C
Cyclamen	(1) Soak for 24 hours in water at 40 °C (2) Sow 2 millimetres deep (3) Temperature of 15–19 °C in dark
Impatiens	(1) Sow on surface of compost (2) Minimum temperature of 21 °C
Meconopsis	(1) 21 days in refrigerator (2) Maximum temperature of 24 °C
Pansy (Viola)	(1) Sow 3 millimetres deep (2) Minimum temperature of 21 °C in dark
Primula	(1) Sow fresh seed on surface of compost (2) Maximum temperature of 20 °C
Sweet pea (Lathyrus)	(1) Chip dark-seeded cultivars on side opposite 'eye' (2) Maximum temperature of 20 °C
Thunbergia	(1) Sow fresh seeds 3 millimetres deep (2) Temperature of 21–24 °C

seed dormancy, and to appreciate the significance of this it is useful to remember the role that the seed has to play, not for a plant growing in our gardens, but for one growing in its natural environment.

A seed encapsulates the new generation. It is the means (in many instances the only means) by which a plant species can survive through periods of adversity. The best and most readily appreciated illustration of this adversity is the temperate climate winter when the temperature is inadequate for most plants to grow and through which annual plants must pass in the form of seed. Clearly, if seeds that were produced and dispersed from the parent plant in the autumn were immediately to germinate and grow, the new generation could very swiftly be obliterated by severe frost and the survival of the whole species would be threatened. The phenomenon of dormancy and/or the presence of substances that temporarily inhibit germination is the means by which survival is ensured. In its simplest but probably commonest manifestation, dormancy is expressed as the inability of seeds to germinate until they have experienced low temperatures for a certain period of time – in effect until they have passed through a winter or the conditions to be expected in a winter. There may be inhibitory chemicals present on the seed that are degraded by cold. Hence the logic behind placing seeds in a refrigerator for a period before sowing them. Sometimes, the chemical factors that confer dormancy do not develop immediately and therein lies the reasoning behind sowing some seeds as soon as they have been collected from the parent plant – clearly, of course, an impossibility with commercially packeted seed.

Less complicated than placing seeds in a refrigerator is to give them the opportunity of experiencing a real winter, and this is much the simplest method to follow with many alpines. Sow the seeds in pans of well-drained sandy compost in the autumn, protect them from slugs and other pests, cover the pans with fine mesh netting to exclude rodents, allow the winter to play its natural part and then bring the seeds into warmth in the spring. The term stratifying is often used for what is essentially this process, although it is usually applied more specifically to the treatment of the fleshy fruits of many trees and shrubs. Layered in sand over winter as I have described, the fleshy outer layers (sometimes containing germination inhibitors) rot away naturally.

Some seeds have exceptionally thick outer coats or testas. These are usually presumed to be an adaptation to a climate with markedly seasonal rainfall. Only when prolonged rain has fallen does the soil become sufficiently saturated to enable an emerged seedling to grow satisfactorily, and only when the soil is in this saturated state does a population of micro-organisms build up able to degrade the seed coat. These are not easy conditions to mimic but the problem can often be circumvented by nicking the seed coat carefully with the tip of a sharp knife on the side opposite the 'eye'.

However, thinking only of a temperate winter as a period of adversity can be misleading, for many garden plants originate in very much warmer climates and, of course, house plants in temperate homes include many sub-tropical and tropical species that would never naturally experience frost at all. For them the period of adversity is a hot, dry period. Placing such seeds in a refrigerator is pointless, and it is high temperatures therefore that are often needed to break dormancy and permit germination to occur. In the few instances of seeds that originate in areas where bush fires are common, germination can actually be induced by covering the seeds with dry straw and setting it aflame.

The necessity for some seeds to be buried and some not is partly related to the likelihood of the young seedling being able to obtain nutrient. Tiny seeds such as those of *Lobelia* have very small nutrient reserves. If buried below the surface, the young seedling would be unable to grow upwards to reach the light and begin photosynthesis. Hence, although some of these small seeds will germinate in the dark, they should always be sown on the surface. A few seeds have adequate nutrient reserves but also an absolute and largely inexplic-

in the summer, some plants display a phenomenon known as high-temperature dormancy – many butterhead lettuce cultivars are notably reluctant to germinate above 25 °C. These often critical requirements and the fact that soil temperatures are tied very closely to the seasons explain why some garden plants are always transplanted after being raised under cover – direct sowing into the open garden would have to be done so late in the year that there would be insufficient time for the plants to mature.

The mechanism by which temperature is related to germination is complex, and although transplanting is one way of minimising the constraints that it causes, it is possible to elevate soil temperatures slightly by using cloches or covering the soil with plastic sheet in advance of sowing. With a transparent plastic mulch, the temperature of the soil surface rises by between 2 and 10 °C; at night the difference is generally between 2 and 4 °C. The temperature differences are similar with black plastic sheet, but with white plastic the soil temperatures are actually reduced as the white surface reflects rather than absorbs heat.

able requirement for light before germination will occur. In contrast, if some seeds (especially those from warm climates) were to germinate on rather than in the soil, the young plant would shrivel away before it had a chance to establish. Some of these actually have light-sensitive inhibitors that are inactivated or degraded once they are buried.

When seeds are sown in seed trays or propagators, the maintenance of a uniform temperature may or may not be easy, depending on the sophistication of the equipment available (Chapter Nine), but it is usually not difficult to maintain a required minimum. Outdoors in the soil of the open garden, temperatures are much less predictable and it is often for this reason that sowing times are closely specified on seed packet directions. Many annual flower seeds and some vegetables (brassicas most notably) will germinate at temperatures as low as 5 or 6 °C (and often over a wide range above this). Others require a very much higher minimum; salvias and cucumbers, for instance, require at least 13 °C. Yet other seeds only germinate well over a very narrow range – celery is particularly fastidious, giving poor results below about 10 °C and above about 19 °C. When soil temperatures become high

The range in size of seeds, even among familiar native plants, is enormous.

The relative effectiveness of two types of plastic cover in elevating the soil temperature preparatory to sowing seeds. Such a temperature rise can result in many days' difference in germination rate and subsequent crop maturity.

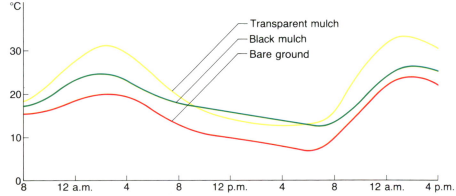

Before I discuss some of the practical aspects of transplanting, I shall mention two means of overcoming the problems of erratic and uncertain germination. The first is by using pelleted seed, that is seed coated with an inert but protective substance such as clay. The real value of pelleted seed, however, is more importantly that of making a small object physically larger and thus more amenable to use in precision sowing machines.

The second is by use of the technique generally called fluid sowing. This is a particularly valuable procedure for it ensures that only already germinated seeds are sown; there will be none of the frustrations associated with gaps in rows or of wide variations in the time of emergence and ultimately of plant size. The fluid component of the technique is a chemically more or less inert jelly which serves to protect the delicate young shoots or roots of the tiny seedling. It is possible to use cellulose wallpaper paste for this purpose, provided that brands are chosen lacking fungicide (which can have an inhibitory effect on growth), but it is simpler to buy a kit containing all of the necessary ingredients and materials. The basic procedure is outlined below.

Perhaps the commonest difficulty for the inexperienced gardener raising plants from seed for the first time is associated with what may be called the weaning process. Even if the gardener has chosen common garden plants, not noted for their difficulty in germination, young seedlings, especially those sown in seed trays rather than directly outdoors, may still die before they have had a chance to establish. (In the open garden, once emerged and given adequate water, a young plant will generally survive.) The need for pricking on is not too difficult to see, for left in the tray where they were sown, the competition between adjacent individual seedlings, the limited supply of nutrients and the usually close atmosphere which can encourage diseases can obviously be detrimental. Provided the young seedlings are handled carefully (preferably by holding not the stem but the young leaves), provided the compost is neither allowed to dry out nor become waterlogged, and provided the young plants are kept fairly cool for a week or so, they should survive and flourish. It is during the next stage that disappointments can arise, for before being transferred from the greenhouse or other sheltered environment to the open garden the

In fluid sowing, the seeds are first germinated on damp tissue in an enclosed container such as a sandwich box. From there, they are carefully washed from the box and trapped on a strainer before being mixed with cellulose paste and sown with an icing bag or plastic bag from which the corner has been cut.

young plants must be hardened off. This is a much misunderstood process but involves the production in the plant of smaller cells with thicker walls; in other words, a structure in which a greater proportion of the volume is of strengthening cell wall material rather than watery protoplasm.

Ultimately, however, all young plants raised and hardened off satisfactorily must be planted in the open garden, and whilst most suffer this experience without difficulty, some types display what is usually called transplanting shock. Among vegetables, tomatoes and cucumbers are notorious in this respect, and even among the brassicas, cauliflowers transplant much less satisfactorily than their close relatives. After transplanting, they may grow very little for a considerable time. This is almost entirely because the roots are inevitably damaged during the transplanting process, take some time to regenerate and thus impair the water uptake of the plant. The larger the plant at transplanting time, the greater will be the check to growth. Always try to transplant vegetables or bedding ornamentals when they are no more than six or seven weeks old although, of course, some plants, like tomatoes, will by then already be fairly large. And ensure that the roots of the plants are never bared – if they are raised in plastic pots always knock out the entire ball of compost or, alternatively, raise the plants in peat pots and thus avoid the need to remove them from the container at all.

There is little evidence to indicate that cutting off some of the leaves of vegetable and other transplants to minimise water loss is beneficial – ensuring that the newly moved plant is watered regularly and the soil maintained in a saturated condition until it is obvious that new growth has commenced is more use. It is also often recommended that transplanting should only be performed in the cool of the evening, but commercial growers routinely transplant vegetable crops on hot days in the height of summer and achieve very successful results simply by providing more or less constant irrigation for several days afterwards. However, should it be necessary to move large shrubs or perennials in

mild periods when they are in full leaf, some cutting back of the top growth is worthwhile, for the root regeneration in such plants is appreciably slower than in annuals. All plants will benefit from a drench application of liquid fertiliser around the roots immediately after transplanting, and this is especially valuable with vegetables and others moved at the height of growth in spring or summer.

So far in this chapter, I have discussed the ways that plants produce seeds and how we as gardeners can make use of and manipulate these natural reproductive processes. Seed production is almost invariably the result of sexual reproduction, although there are a few species of plant that are able to produce seeds without fertilisation by a process called apomixis. But many plants also reproduce vegetatively; they form offspring simply by growing large and shedding parts of themselves. All gardeners will be familiar with numerous examples of vegetative reproduction – the rhizomes of couch grass, the small 'daughter' cormlets attached to gladiolus corms, the multiplication of daffodil bulbs, the forming of runners by strawberry plants and many, many more. In Chapter Two I mentioned that these structures are modified stems or roots. In many ways, vegetative reproduction is extremely important in gardens. In one form or another it is the most important method of spread for some of the most pernicious weeds like horsetail, couch grass, ground elder and oxalis. But it is also the means by which many of the most important garden plants are perpetuated artificially. Vegetative propagation of plants is used for three main reasons.

In many instances, such as the numerous corm- and bulb-forming ornamentals, the time required from seed sowing to plant maturity and flowering is several years, yet by planting the corm or bulb itself, the end result is obtained in the first season. It makes sense, therefore, to start with bulbs or corms wherever possible and then allow those types willing and able to self-seed to build up an extending population in the garden. But of rather more widespread significance is the fact that in many plants the ability actually to produce seed at all has been lost or, alternatively, that the plant has been so bred and selected that it will only produce offspring of uncertain genetic constitution. Many dahlias and roses are typical of the first category; their much admired double flowers no longer bear stamens. Many types of carnation and chrysanthemum, potatoes and most fruit trees and bushes are common instances of the second type. Indeed, although it is common practice to refer to the many named types of rose, potato and apple ('Peace', 'Maris Piper' and 'Discovery', for instance) as cultivars, they are strictly clones – populations of individuals of identical genetic constitution. They are perpetuated without sexual reproduction and there is no opportunity therefore for the genetic material to be rearranged. This has its advantages and disadvantages.

A clone will certainly be uniform in appearance and performance under defined conditions, but the continual transfer of vegetative tissues from one generation to the next means that any abnormalities or defects will be transferred too. Most importantly, disease-causing viruses which more or less permeate the tissues will be passed from parent to offspring (although some viruses can be transferred through the seeds, this is generally a much less common and less serious occurrence). It is no coincidence that many of the cultivated plants most prone to virus problems are vegetatively propagated – potatoes, carnations, dahlias and fruit trees and bushes especially.

In the past, before the nature and biology of viruses was understood, gardeners came to expect and accept a gradual deterioration in the quality of their plants, but several techniques are now used to overcome the difficulties and stock plants called virus-free or certified are offered for sale. In some instances these stocks are maintained by avoiding the viruses – British potato 'seed' tubers for instance are raised in Scotland, Northern Ireland or other regions where there are few virus-carrying aphids. But when these plants are grown in gardens in the south of Britain, the aphids and their viruses will soon arrive and gardeners in these regions who save their own tubers for replanting must expect a gradual decline in quality within a few years (see p. 140). With many other types of plant, ranging from orchids to rhubarb, modern techniques of micropropagation (p. 132) offer the opportunity for removing and propagating only those cells known from careful microscopical and biochemical study to be free from virus.

In using bulbs, corms, tubers, rhizomes, offsets or runners, we are making use of natural reproductive processes and adapting them to our own ends. But the ability of plants to regenerate tissues means that we can sometimes adapt natural processes much more significantly and in some instances devise

The structure of a bulb, a greatly abbreviated and telescoped shoot, is only apparent when it is sliced vertically. The stem is the plate-like body at the base, its true nature revealed by the fact that this is where the roots arise.

operations that really have little likelihood of ever occurring naturally. Among these techniques are the taking of root, stem or leaf cuttings, and the induction of root formation by layering or air layering. Examples of these are shown overleaf. The principles in each case are roughly the same. Either solely through the natural response to the stimulus of wounding or with the additional stimulus of an artificial hormone-like chemical such as 1-naphthylacetic acid, roots form and new plants thus arise. But there is a further procedure, wholly artificial, that has come to assume an extremely significant role in horticulture.

Grafting and its abbreviated cousin, budding, involve the insertion of a large or small part of a shoot (generally called a scion) into the tissues of an already established plant, generally called a rootstock. There is fusion between the two plants to produce a viable individual, but the upper part remains genetically distinct from the roots. The techniques of grafting and budding have been known since ancient times and they are important for several reasons. Like the taking of cuttings, they offer the opportunity of multiplying plants that produce no seeds. But they are quicker and more efficient than cuttings because they allow the use of an already established root system and thus a swifter progression to a flowering or fruiting plant and they offer (budding especially) to the commercial grower the facility for producing a very large number of marketable plants from very little source material. Unlike cuttings, grafting enables the multiplication not only of plants that produce no seeds but also of plants that produce weak roots too. There are drawbacks, however, and there are problems. There is always likely to be shoot growth (suckers) from the rootstock as well as from the scion and this is a perennial problem for rose growers. Moreover, there must be close compatibility between the two plants – their tissues must be capable of uniting. But not only should the two tissues of the two plants be compatible, they should be of similar growth rate too, for an ostensibly ornamental tree with a trunk growing at twice the

speed of its crown produces a bizarre and ugly combination. Where such matching is impossible, a basal graft with the union at ground level rather than a crown graft with the union 2 or 3 metres up the stem will minimise the ugliness. I have shown overleaf some of the commonest and most useful types of grafting and budding that should be within the capability of most gardeners.

Grafting, as with this crown graft, is undertaken for a wide variety of reasons, but most commonly because it shortens the time before a mature plant is obtained.

Grafting and budding techniques.

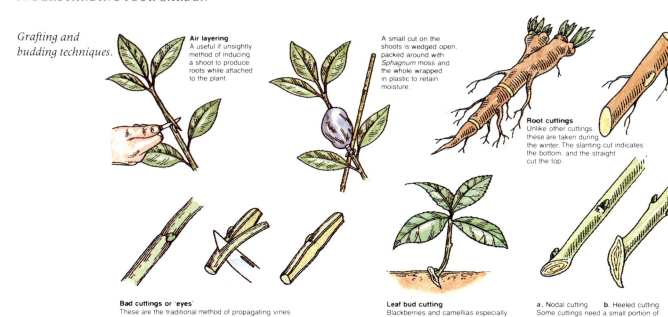

Air layering
A useful if unsightly method of inducing a shoot to produce roots while attached to the plant.

A small cut on the shoots is wedged open, packed around with *Sphagnum* moss and the whole wrapped in plastic to retain moisture.

Root cuttings
Unlike other cuttings, these are taken during the winter. The slanting cut indicates the bottom, and the straight cut the top.

Bad cuttings or 'eyes'
These are the traditional method of propagating vines but the exposed tissue must be kept moist.

Leaf bud cutting
Blackberries and camellias especially are most simply propagated in this way; the bud should be half buried.

a. Nodal cutting   **b.** Heeled cutting
Some cuttings need a small portion of old wood, the heel, to root successfully.

Whip and tongue graft
Often used on young fruit tree stocks where the rootstock (below) and scion (above) are of similar thickness.

Side graft
A useful method for rejuvenating a tree by inserting a new scion into an old rootstock framework.

Cleft graft
Used where the rootstock is of much greater thickness than the scion. A simple method but leaving an open wound.

Splice graft
Similar to the whip and tongue technique but simpler and relying on secure binding for success.

Crown graft
More sophisticated than cleft grafting and often used to rejuvenate worn out fruit trees.

Budding
A bud is a small graft that makes use of one bud only and is thus very economical as plant material.

1. A small shallow 'T' shaped slit should be cut in the bark of the stock and gently lifted with the tip of the knife.

2. The bark sliver bearing the single bud should be carefully inserted into the slit, ensuring that the bud itself faces upwards.

3. Binding is as important with budding as with other forms of grafting to hold the tissues firmly and create good adhesion.

Not only did the Romans practise grafting, they also knew of one of its most important yet unexpected benefits. In many cases, the rootstock exercises an influence over the growth of the scion and two identical scions, from the same stock plant may result in very differently sized trees if they are grafted onto two different types of rootstocks. The mechanism of this phenomenon is hormone-mediated, and it has led within the present century to a revolution in the production of fruit trees. First at the old John Innes Horticultural Institution at Merton and then at the East Malling Research Station in Kent, programmes were commenced to breed apple rootstocks that would exercise a defined, predictable and especially a dwarfing effect on any apple cultivars (or clones) grafted onto them. The rootstocks were numbered in two main series, the Malling–Merton or MM series and the later Malling or M series. Many different stocks were bred although relatively few have become available or indeed are of value to gardeners. The effects of the most widely used of them are indicated below, but certain general points should be made. Usually, the more dwarfing the effect of the rootstock, the better the soil and other growing conditions required, and the very dwarfing M.27 is really at its best in most gardens when used with a John Innes No. 3 potting compost in a large container such as a wooden half-barrel. M.27 is also unsuitable for apples that are of weak constitution such as 'Rev. W. Wilks' or 'Lane's Prince Albert', for the additional reduction in vigour that it imparts

results in a very feeble plant indeed. However M.27 is valuable for naturally vigorous types such as 'Blenheim Orange', 'Bramley's Seedling' or 'Crispin'. Trees grown on any of the more dwarfing rootstocks require staking throughout their lives. There is much less choice in rootstock type with other tree fruits, but I have summarised above the characteristics of those that are available.

Important fruit tree rootstocks for garden use and their effects on tree size

Fruit	Rootstock	Characteristics
Apple	M.27	Very dwarfing (tree attains 1.5 metres); only suitable for very good soils; requires permanent stake
	M.9	Dwarfing (tree attains 2 metres); the best rootstock for general garden use; requires permanent stake
	M.26	Semi-dwarfing (tree attains 2.5 metres); good for poorer soils
	MM.106	Semi-vigorous (tree attains 3–4 metres); better than M.26 for weaker growing cultivars
Pear	Quince C	Moderately dwarfing (tree attains 3–5 metres); best for general garden use but not for poor soils
	Quince A	Moderately vigorous (tree attains 4–5 metres); best for poorer soils
Plum	Pixy	Dwarfing (tree attains 2–3 metres but must be pruned to achieve this restriction)
	St Julien A	Semi-dwarfing (tree attains 3 metres – but must be pruned to achieve this restriction)
Cherry	Colt	Semi-vigorous (tree attains 3 metres – but must be pruned to achieve this restriction)
	F 12/1	Vigorous (tree attains 10 metres)
Damson	Pixy	See plum
	St Julien A	See plum
Peach	St Julien A	See plum
Quince	Own roots	
Medlar	Quince	

The relative growth rates after twelve years of the same apple variety when grafted onto four of the most commonly used rootstocks. The tallest tree shown is 3.5 metres tall. In this way, the size of tree can be matched to the area available.

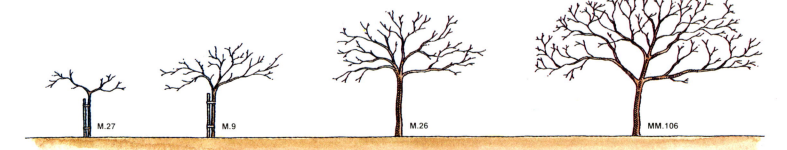

M.27   M.9   M.26   MM.106

Modern dwarfing rootstocks and new, carefully bred varieties mean that productive apple growing is now possible in almost every garden, no matter where it is situated or what its size.

The traditional, natural ways of growing plants from seed have been replaced in some instances by artificial culture, not from seeds but from small pieces of tissue. This is a development that accelerates the reproductive process greatly and enables many plants of identical genetic type to be produced.

Finally, a look at the present and also the future of plant propagation for, in commerce at least, many of the traditional techniques have already been left behind. Three expressions have found their way into everyday language in recent years and are likely to be heard of increasingly as time goes by – micropropagation, tissue culture and genetic engineering. The last-mentioned, genetic engineering, will always be a technique beyond gardeners' reach, entailing such operations as the micro-dissection or chemical alteration of living cells and the removal from them of parts of chromosomes bearing genes for transplanting into other cells. I have already mentioned micropropagation in relation to the production of plants free from virus contamination and it goes hand in hand with tissue culture. It was discovered many years ago that, with care, it is possible to regenerate entire plants from very small pieces of tissue. Micropropagation is this technique – the multiplication of plants by using a very small portion of tissue, containing only a few cells as the starting point or 'initial'. These tiny pieces are usually removed from zones of meristematic tissue such as root or shoot tips, and the whole delicate operation must therefore be performed under a microscope in a laboratory. And, because of the extreme fragility of the tissue pieces and their vulnerability to contaminating bacteria or fungi, it must also be done under sterile conditions. Clearly, after taking such precautions, it would be pointless placing the tiny fragments into soil or compost, and so they are usually transferred to a sterile jelly containing a carefully balanced blend of nutrients and they are incubated for several weeks. Gradually, the tiny fragments begin to form roots and leaves and become recognisable as young plants. Eventually, as they become large enough to handle easily, they are transferred to sterile seedling compost and ultimately become as robust as any other transplants. The first micropropagation kits have now become available to gardeners, although the technique is still far from simple for anyone not having access to sterile growing conditions.

CHAPTER EIGHT

Garden problems – pests, diseases and weeds

The garden free from pests, diseases and weeds is the garden of an unattainable Utopia. Nothing so reminds us that we are but one factor in a whole complex of ecological influences as when our plants suffer at the expense of some other organism, be it animal, vegetable, fungal, bacterial or viral. But nor does anything so frequently give cause for unwarranted alarm or serve as a justification for some wholly unnecessary expenditure of money and effort on chemical treatments. Despite the fact that gardeners are now barraged with advice and literature (especially from chemical manufacturers), I am still surprised and saddened at the number who fail to appreciate the distinction between the serious and not so serious problem, and who fail to realise that each chemical pesticide has a rather specific task to accomplish. With weeds, just as with insects and other pests and with fungal, bacterial and viral diseases, a little learning can be a valuable thing. A sense of proportion is needed in deciding first, if a particular problem actually justifies any control measure at all and second, if so, which of the various options is appropriate.

The detailed treatment of each and every garden pest, disease and weed problem is beyond my scope here, but I hope that I can convey certain general principles of their biology and treatment. You should thus be equipped to make more meaningful decisions on the best course of action to adopt in any particular situation. But I shall first describe the main types of organism with which we are concerned and outline the various ways that they live and feed. Understanding this will help to show the distinction between the very damaging and the slightly damaging on the one hand, and the insignificant and the beneficial on the other. It will also make more logical the many preventative measures that can be taken to lessen the likelihood or impact of any damage. And if direct control measures do prove necessary, it will enable them to be planned meaningfully.

Although the various types of harmful or competing organism are generally grouped together (as

The attractive, productive garden is not an accident, but only comes about after careful attention has been paid to pest and disease control. The unnatural environment of a garden actually tends to encourage the natural enemies of plants to proliferate.

they have been in this chapter), the ways that they bring about their several effects vary and it makes sound sense therefore to consider separately the three groupings of pests, diseases and weeds.

Pests are animals that through their activities (usually their feeding activities) work counter to our objectives. They are not intrinsically any different from other animals whose lives never come to our attention, and we must never forget that an animal is designated a pest solely in reference to some human activity or purpose. This definition may give us a sound reason for reacting in a particular way to that animal's presence or behaviour; it does not give us a right to do so.

Almost every subdivision of the animal kingdom likely to be found in a garden has representatives that can be considered pests. This is not surprising, for a very large proportion of animals are herbivorous – they eat plants – and the majority of plants in gardens are there because we so choose. The largest groups of animals generally contain the largest number of garden-inhabiting species and also the largest number of pests. The main groups of garden pests are summarised on the right. It will be seen that they range from nematodes at the lower end of the complexity spectrum to mammals at the other. The numerical importance of arthropods in general and insects in particular in the animal kingdom is well known, and this is very much borne out by the number of insect pests that gardeners encounter.

Of course, these many different types of organism range vastly in their structural complexity, evolutionary advancement and behavioural sophistication. Nonetheless, there are certain features common to all animals and four of them go a long way towards explaining how any particular one becomes a pest, how it achieves its ends and also indicates how we might combat it. These four factors are where it lives, how numerous it is, how it reproduces and in which of two basic ways it feeds.

Whilst many animals range through several different habitats, most tend to live predominantly *in*

Principal groups of animal pests (and their relative numbers) to be found in gardens

Animal group	Approximate number of important British garden pest species
Eelworms (Nematoda)	30
Earthworms (Annelida)	1
Slugs, snails (Mollusca)	8
Arthropods (Arthropoda)	
Symphylids (Symphyla)	1
Springtails (Collembola)	5
Woodlice (Crustacea)	6
Insects (Insecta)	
Grasshoppers, crickets (Saltatoria)	3
Leaf-hoppers, capsid bugs, mealy bugs, scale insects, whiteflies and allied groups (Hemiptera)	60
Earwigs (Dermaptera)	1
Thrips (Thysanoptera)	8
Aphids and allied groups (Homoptera)	75
Beetles (Coleoptera)	50
Flies (Diptera)	35
Butterflies and moths (Lepidoptera)	55
Wasps, bees, ants, sawflies (Hymenoptera)	35
Mites (Arachnida)	30
Millepedes (Myriapoda)	5
Birds (Aves)	11
Mammals (Mammalia)	12

Chewing pests, such as caterpillars, are usually among those most easily controlled. They are often relatively conspicuous and generally present in small numbers.

the soil or predominantly *above* the soil either for their entire lives or for clearly defined developmental stages. Earthworms, for instance, are exclusively soil inhabiting and moles predominantly so. Bullfinches live exclusively above the soil, as do large white butterflies, squirrels and many species of wasp. Crane flies and winter moths, however, have distinct phases – the crane fly larva or leatherjacket is subterranean and so is its pupa, but the adult is wholly airborne. The winter moth spends a period of its life in the soil as a pupa, but the larvae and adults live above ground. There are nonetheless some animals – voles and many adult beetles for instance – that routinely spend part of their time as adults in the soil but also (very often during the night) emerge above it. The above-ground period of animal activity may be spent on the surface of the soil (many beetles), on the surface of plants (many caterpillars) or within plant tissues (many fly larvae). I attach especial importance to the subterranean/above-ground distinction however.

Anyone who has ever tried to catch a mole will realise the special benefits that are conferred by a subterranean existence. The soil is a dark, opaque, unyielding medium that hides like nothing else hides. Inability to see your quarry need not of course be a great hindrance to controlling it (bacteria are invisible to the naked eye although sore throats are not too difficult to treat), but the soil offers special problems to the would-be pest trapper (and the fungal disease trapper too). For instance, many pesticidal chemicals have fairly large molecules and often fairly low volatility, neither attribute aiding their movement through the soil. And the soil, to put it bluntly, is very big. The quantities of chemical needed to permeate fully even a small bed or border of soil are very great. A pest such as an eelworm that lives permanently underground will always be intrinsically very difficult to control by any direct approach method.

Some pests hide out of sight not in the soil but actually within the plant tissues on which they feed. I have already mentioned in this respect fly larvae (the cabbage root fly, for instance), but many

The old adage 'know your enemy before you take up arms against him', is well applied to the wasp. Though its nuisance factor may be great, the wasp's activities as a pest are in fact minimal and it plays a valuable role for much of the year in catching other insects on which to feed its young.

caterpillars too (the pea moth, for example), beetles (the larvae of the raspberry beetle are notable examples) and earwigs among numerous others are generally invisible until their feeding activities reveal their presence. Whilst plant tissue may not be as unyielding as soil, it creates different problems for it still offers a barrier against chemical diffusion and, of course, any chemical directed against the pest must at the same time be harmless to the plant host.

The second of the attributes of pests to which I attach special importance is the numbers in which they occur, although this is not an invariable indicator of the amount of trouble they will cause. The numbers are generally related in some way to the size of the animals – a given habitat, be it garden or anything else, will offer sufficient food to support few large or many small creatures. You may, for instance, have only one rabbit in your garden, yet it will cause you as much anxiety as several hundred caterpillars or several thousand aphids.

Third, we should consider the way in which the pest reproduces. Does it, in the manner of slugs, snails and starlings, lay eggs which hatch directly into juveniles that are stucturally like their parents but require varying periods of time before attaining maturity? Or do the eggs pass, as with moths, butterflies and flies, through one or more larval or other immature stages, perhaps finally undergoing a complete pupal or other metamorphosis before becoming adult? Alternatively, perhaps the pest is like the rabbit, the mole or, more surprisingly, many types of aphid, and gives birth to living young. The relatively rapid reproduction rate of rabbits and their direct production of living off-spring stands them in good stead among other animals and contributes to their overall national pest status, but within any one garden it is more or less irrelevant. It is their large individual size that matters here. But the tiny aphid, which on its own is insignificant, is elevated to major prominence very largely because it can attain maturity within about seven days and can produce several hundred offspring without the need for fertilisation. The scarcely credible statistic that one adult aphid could give rise to millions of tonnes of descendants within a single season is easily proved mathematically – although, as I shall explain later (p. 144), it could never be realised in practice.

The final feature to bear in mind in designating the seriousness of a pest is its method of feeding and I find it useful to divide the animal kingdom into the two categories of chewers and suckers. The

The gooseberry sawfly larva is well camouflaged and consumes plant matter at a prodigious rate. This underlines the need for vigilance in order to spot the early signs of pest attack before major harm can be done.

Scale insects are very difficult to control for their protective covering shields them against pesticide sprays. Such sap-feeding creatures can often only be combated by systemic pesticides that actually permeate the sap.

chewers comprise by far the larger of the two groups and although their chewing organs may range from the conventional teeth of a vole through the horny mandibles of a beetle to the rasping tongue or radula of a slug, the end result is the same. Pieces of plant tissue are removed wholesale, and damage to the plant is actually manifest in one of two ways. Either its food-manufacturing or other functions are impaired because of the loss of photosynthetic or other tissues, or the damage caused admits decay-forming bacteria and fungi. The former effect is usually the more important on leaves, the latter on flowers, fruits and roots.

The species of sucking animals are numerically many fewer than the chewers, but the very nature of their feeding method ensures that they are all potentially significant as pests. They are all insects and they all feed on plant sap by inserting modified, hypodermic-like mouthparts into the phloem. The group includes aphids, capsid bugs, frog-hoppers, lace bugs, leaf-hoppers, mealy bugs, scale insects, thrips and whiteflies – all names extremely familiar to anyone who gardens. They achieve their purpose of withdrawing sap at the expense of weakening the plant which is denied the nutrients that the sap contains. But many species of sucking pest have a separate and incidental adverse effect for they provide the means by which viruses are transmitted from one plant to another, a feature that I shall consider in more detail later.

I hope it will be evident that it is possible by considering various combinations of these major attributes to obtain a formula or recipe for the likelihood of an organism attaining serious pest status. The most serious pest is likely to have a preference for garden plants over wild plants as food and to be very large individually, to have a very rapid reproduction rate or to have some particular feature (a subterranean habit, for instance) that renders it especially inaccessible to potential control measures.

However, one factor common to almost all pest organisms offers some scope for control. There are very few plants in active growth in the garden during the winter in temperate countries and in consequence there is little food for pests. Some may move back into natural habitats to seek food there, but most enter a more or less dormant phase. This may take one of three forms – the pests may metamorphose to a pupa or similar body, the adults may die out leaving only eggs, or the mature creatures may themselves become torpid or sluggish and enter on a hibernation. Whichever method is adopted, the winter offers to gardeners a breathing space when their plants are fairly safe from attack and the pests are relatively static and vulnerable. Thus the gardener can adopt some control measures that may not be feasible during the rest of the year.

Some of the factors that I have outlined in relation to pests (especially the importance of the soil as a protective environment) are applicable to disease-causing organisms too, although the three groups of fungi, bacteria and viruses are themselves very different from each other biologically. Overall, fungi are the most important of the three in gardens, although individual types of bacteria and virus are very significant. The fungi are a highly diverse group of organisms, ranging from the familiar mushrooms and toadstools to mildews, rusts and numerous obscure and elusive microscopic groups, many only distantly related to each other. Numerically, the group is very large – there are probably upwards of 5000 genera and 100 000 species – and there may in fact be more types of fungi than of flowering plants in the world. They all share, however, a basic thread-like structural unit called a hypha, reproduction methods based on spores rather than seeds or other bodies, and a lack of chlorophyll and hence an inability to photosynthesise. It is from their lack of photosynthesis and their dependence on other organisms as sources of nutrient that fungi derive their status as pathogens – causes of disease. But there is a significant division of fungi on a nutritional basis that to a considerable extent cuts across taxonomic boundaries and has a very practical importance. For whilst some fungi can obtain nutrient only from dead organic matter and some only from living organisms, yet others have the versatility to feed on living organisms and then continue to live on their dead remains when they have succumbed to the attack. Fungi that can feed on dead matter are called saprotrophs, those that can attack living tissues, biotrophs, with the further designation of obligate or non-obligate to indicate the relative exclusiveness of this status.

There are relatively few obligate biotrophs among fungi in gardens – rusts and mildews are the best known and most important. It is to their advantage that the affected host plants remain alive for as long as possible and they have sophisticated and modified hyphae that penetrate living cells to withdraw nutrient while causing minimal direct damage. In time, of course, there will be a gradual weakening of the plant just as there is when sucking pests remove the sap. Obligate saprotrophs, by definition, will of course only be adversely important in attacking non-living structural components of the garden – fence posts and wooden tubs for example – although they are of immensely beneficial significance in aiding the breakdown of compost and dead material in the soil and environment generally. By far the largest group of disease-causing garden fungi are the non-obligate biotrophs and saprotrophs. These function either by attacking dead parts of plants, such as frost-damaged twigs, and then spreading into living tissue, or by attacking particularly soft and vulnerable parts of living plants, such as their flowers and fruit, and then continuing to feed on the dead remains.

When a plant succumbs to fungal attack from an obligate biotroph, the fungus must have some means of moving on to another food source or of surviving for a period with no nutrient at all. Even a non-obligate biotroph will be faced with problems at the end of the season, much in the same way as an insect or other pest – partly because of a lack of nutrient, but also because falling temperatures and the onset of frosts render further growth difficult. Usually, fungi form specialised spores or other bodies able to survive these periods of adversity but, as with pests, such times offer to the gardener an opportunity to impose control measures, either by breaking a phase in the organism's life cycle or by diminishing the overall numbers of spores or other structures before the start of the new season.

Because fungi, like plants but unlike animals, lack any powers of locomotion, they are dependent on some other agency for their spread from place to place. Far and away the most important means of fungal dispersal is by utilising the wind to carry their very light microscopic spores. In this way, not only can plant-to-plant spread occur, but spores of pathogenic fungi may be carried aloft into the atmosphere in their countless millions, subsequently to bring about plant diseases many hundreds of kilometres from their source. However, on a localised level much spread of fungal diseases in gardens is due to rain splash (either of spores or of contaminated soil) or to the inadvertent or careless transfer

Bacteria are rarely the cause of major garden diseases, but some of their effects are immediately apparent – for instance, the foul-smelling symptoms of bacterial soft rot on vegetables and fruit.

of contaminated soil or diseased plants by gardeners themselves.

I have mentioned the soil several times in relation to fungi and it does have a rather special role both for fungal biology and, in consequence, for their control. Almost every species of fungus spends at least part of its life cycle in the soil and thus derives much the same benefits from the protection the soil can give as those I outlined for pests on p. 135. However, many fungi can and do survive naturally in the soil only for very short periods (often in the form of spores) before they succumb to natural enemies (see p. 146), and some disease control measures utilise this fact. But some of the most difficult types of disease to control are those caused by fungi that live permanently in the soil from where they launch their attacks – usually against plant roots.

There are many fewer species of bacteria (probably about 1600) than there are of fungi, although they make up for this shortage in variety by producing quite astronomically large numbers of individuals. All are microscopic and most are single-celled. Bacteria occur in almost every environment and habitat on earth and are especially abundant in the soil – many thousands of millions of bacterial cells may be present in every gram of soil. Relatively few produce spores and most reproduce by simple fission – each cell divides into two. Fortunately, relatively few species of bacteria are significant pathogens, either to plants or animals, and in gardens the greatest importance probably attaches to the so-called spoiling types – those that attack stored produce such as fruit, bulbs or tubers and bring about a disintegration of the tissues and the effect known as soft rot.

Viruses are almost totally different from all other organisms and are sometimes described as self-replicating chemicals rather than living things. Structurally they are extremely small and can be rendered apparent only under the high magnifications possible in electron microscopy. (This is a technique that utilises a very-short-wavelength beam of electrons rather than light as the illuminating source – the ability to resolve very small objects is a direct function of the wavelength of the illuminating source and the maximum usable magnification possible with a light microscope is about 1400 times). Structurally, viruses are very simple, comprising a nucleic acid and one or more proteins. They can exist only in living cells and usurp the subcellular structures of their host to their own ends. They usually occur more or less systemically (throughout the tissues) of an affected plant and are not confined to the region in which the symptoms appear. This results in the transmission of the virus contamination from generation to generation of vegetatively propagated plants such as potatoes, chrysanthemums, carnations and many fruit trees and bushes, and explains why these plants are so often subject to virus problems. Viruses are also spread naturally from parent to daughter plants in a

Though visually attractive, this wallflower betrays its infection by a virus which has brought about the yellow streaking on its dark red petals.

comparable manner, but most also have some facility for being more widely dispersed. Having no specific dispersal bodies such as spores, however, viruses are dependent either on mechanisms for the natural spread of the parts of the host plants in which they occur (seeds especially), or on some other organism that will carry virus-contaminated sap from one plant to another. Almost every type of organism that feeds on or comes into contact with plant sap is implicated in the transfer of at least some type of virus. In temperate climates, such as that of Britain, aphids are the most significant air-borne virus vectors. Eelworms are the most important in the soil – where, of course, the spatial movment is much less and viruses are spread from plant to plant much more slowly.

Unlike pests, disease-causing organisms are rarely seen – most fungi and of course all bacteria and viruses are microscopic – and it is their effects on the plant (the symptoms of attack) that must provide the clues to the most likely control measures. Whilst some effects are peculiar to certain types of disease-causing organism, others have a wide range of causal factors. It is for the latter cases that control measures can prove particularly difficult – different types of pathogen respond to different types of control measure.

Weeds differ most greatly from the other organisms that adversely affect gardening activities in that they do not directly attack our plants nor necessarily bring about their demise. As plants themselves, however, they are in direct competition with cultivated vegetation for the same basic essentials of light, water, nutrients and, implicit in all of these, for the space in which to grow. Gardeners' feelings about weeds are embodied in terminology, and whilst the speedwell that grows in the beds, borders and the lawn is a weed to be eradicated, the same plant in the fields and hedgerows is an attractive wild flower to be conserved. Weeds are the native vegetation of an area, whereas garden plants are usually exotic, alien and have been subject to some intensive artificial breeding and selection processes. The consequence is that

Speedwells are attractive when regarded as wild flowers but some species, especially the creeping species that invades lawns, can be highly intrusive weeds. There is no fully effective treatment for lawn speedwell.

whilst the cultivated plants may be more attractive and have larger flowers or more fruit, the weeds are ecologically more fitted to the environment of your garden and are almost always aggressively competitive.

It is worth bearing in mind that successful weeds have certain attributes that set them apart not just from garden plants but also from many other native species. And depending on which of these attributes is important in each instance, you should have some indication of those control measures that are likely to succeed and those that are not.

First I shall deal with annual weeds, those that produce flowers and set seed within the course of a single season. The seeds of some annual weed species require no period of cold or enforced dormancy before they can germinate and if, as in groundsel (*Senecio vulgaris*), this is combined with the ability to produce flowers and seeds very early in life, more than one generation may occur within the course of a single season. Even weed seeds that do undergo a dormant phase may vary in the time of year when subsequently they are able to germinate – some like those of the red dead-nettle (*Lamium purpureum*) can germinate only in the spring, whilst those of the annual nettle (*Urtica urens*) can germinate both in spring and in autumn. Generally, the longer the period of the year over which its seeds are able to germinate, the more successful a weed is likely to be. A few species (shepherd's purse, *Capsella bursa-pastoris*, and annual meadow grass, *Poa annua*, for instance) can actually germinate in every month of the year.

A second feature of importance for annual weeds is the number of seeds that they produce. In general, the greater the number of seeds, the greater the likelihood of some escaping the attentions of mammals, birds and other animals and, of course, of decay-causing micro-organisms – and hence the greater the chance of the weed attaining significance in the garden. The table overleaf indicates the average number of seeds produced by a range of common annual weeds, but these figures should be taken in conjunction with other factors.

The sow thistle, for instance, produces a very large number of seeds, but anyone who has observed sow thistles will know that they are a magnet for finches and other seed-eating birds and so the actual proportion of seeds reaching the soil may be relatively small.

Once in the soil, the seeds of most annual weeds have the potential to survive for at least ten years – although not all of course manage this for the various reasons that I have outlined above. Seeds of many weed species can survive for much longer periods moreover, and in an unweeded garden where new plants are growing and adding their seeds year by year, the numbers actually present in the soil can reach formidable proportions. The highest number of seeds actually recovered from a sample of soil is about 90 000 per square metre, but it is possible to estimate the number of seeds present in an area of your garden by using the reasonable approximation of 10 per cent germinating in one year. Count the number of weed seedlings emerging in a defined area (say, half a square metre) and multiply this by ten. Extrapolating further, you should be able to determine from a knowledge of the longevity of the particular weed, how many years must elapse before the population of that species will vanish from your garden – provided, of course, that you prevent any of each year's emerging plants from seeding and adding to the total (see p. 148) and also providing that the species does not have freely air-borne seeds that are likely to be blown in significantly from nearby gardens.

There is one further feature of the seeding of annual weeds that has a considerable bearing on its efficiency and on the merit of particular control measures. Groundsel and annual meadow grass are examples of weeds that, in addition to all their other attributes, can continue to mature their seeds if they are hoed down during flowering. If the weather and soil are moist at the time, they may actually produce new roots in response to the wounding (see p. 148) and end up, in effect, merely having been transplanted.

Approximate number of seeds produced each year by some common British annual garden weeds

Species	Common name	Approximate average number of seeds per plant
Aethusa cynapium	Fool's parsley	6000
Capsella bursa-pastoris	Shepherd's purse	4000
Cardamine hirsuta	Hairy bitter-cress	600
Chenopodium album	Fat hen	3000
Euphorbia peplus	Petty spurge	250
Fumaria officinalis	Fumitory	less than 100
Galium aparine	Cleavers	less than 100
Lapsana communis	Nipplewort	1000
Matricaria matricarioides	Rayless mayweed	7000
Myosotis arvensis	Field forget-me-not	3000
Papaver rhoeas	Field poppy	17000
Senecio vulgaris	Groundsel	1000
Solanum nigrum	Black nightshade	10000
Sonchus spp.	Sow thistles	18000
Stellaria media	Chickweed	2500
Thlaspi arvense	Penny cress	2000
Urtica urens	Annual nettle	1000
Veronica spp.	Speedwells	2000

Approximate longevity in soil of the seeds of some common British weed species[a]

Species	Common name	Approximate longevity in soil (years)
Aethusa cynapium	Fool's parsley	10
Calystegia sepium	Greater bindweed	50
Capsella bursa-pastoris	Shepherd's purse	30
Chenopodium album	Fat hen	30
Cirsium arvense	Creeping thistle	23
Fumaria officinalis	Fumitory	30
Papaver rhoeas	Field poppy	80
Plantago lanceolata	Plantain	15
Plantago major	Broad-leaved plantain	30
Polygonum persicaria	Persicaria	35
Rumex crispus	Curly dock	45
Rumex obtusifolius	Broad-leaved dock	45
Solanum nigrum	Black nightshade	40
Stellaria media	Chickweed	17
Thlaspi arvense	Penny cress	40
Veronica spp.	Speedwells	10

[a] not all of these species are commonly troublesome in gardens

Most lawn weeds, such as these plantains, achieve their resilience through a rosette form with greatly shortened stems and a growing point below the level at which mower blades operate.

Methods of multiplication and spread of some important British perennial garden weeds

Species	Common name	Method(s) of multiplication and spread
Achillea millefolium	Yarrow	Rhizomes, some seeds
Aegopodium podagraria	Ground elder	Rhizomes, some seeds
Agropyron repens	Couch grass	Rhizomes, few seeds
Bellis perennis	Daisy	Seeds, short creeping stems
Calystegia sepium	Hedge bindweed	Rhizomes, creeping stems
Cirsium arvense	Creeping thistle	Creeping roots
Convolvulus arvensis	Field bindweed	Creeping roots, some seeds
Epilobium angustifolium	Rosebay willowherb	Seeds
Epilobium montanum	Broad-leaved willowherb	Seeds
Equisetum arvense	Field horsetail	Rhizomes
Lamium album	White dead-nettle	Rhizomes
Oxalis spp.	Pink-flowered oxalis	Bulbils
Ranunculus ficaria	Lesser celandine	Bulbils, seeds, root tubers
Ranunculus repens	Creeping buttercup	Creeping stems, some seeds
Reynoutria japonica	Japanese knotweed	Rhizomes, some seeds
Sagina spp.	Pearlworts	Seeds
Taraxacum officinale	Dandelion	Seeds and pieces of root
Trifolium repens	White clover	Creeping stems, seeds
Urtica dioica	Stinging nettle	Creeping stems, seeds
Veronica filiformis	Creeping speedwell	Creeping stems

For perennial weed species, seeds are rarely of great significance as part of the perennating process or spread in gardens. Such weeds are almost invariably dependent for their success on some method of vegetative growth – deeply growing or creeping roots, far-spreading rhizomes, creeping stems or bulbils are the most important (see left). Rapid growth of a root or rhizome to great length provides a straightforward difficulty in that there is physically a large amount of plant to be removed – and, when growing deeply in the soil, there is the added problem of actually locating all of the growth. But this purely physical problem is of small account compared with the ability of the plant to regenerate from small parts of stem or root that may be left behind. They vary in this ability – dandelion (*Taraxacum officinale*), for example, can regenerate from any part of its tap root whereas docks (*Rumex* spp.), can only do so from the top 7–10 centimetres. Plants such as couch grass (*Agropyron repens*), ground elder, (*Aegopodium podagraria*), or horsetail (*Equisetum arvense*), with creeping rhizomes present special problems for they have frequent nodes with buds at each. If the rhizome is severed or otherwise disturbed, these buds are stimulated to grow and when, as is often the case, the whole structure is innately brittle, the physical eradication of the plant becomes an exceedingly difficult task.

Rather fewer important weeds have creeping roots, but these too can regenerate when they are severed or damaged. Creeping thistle, (*Cirsium arvense*) is probably the most significant garden weed in this category. But perhaps in a league of their own are the weeds that produce bulbils (tiny bulbs) for these are ineradicable by physical means and when, as in the pink-flowered *Oxalis* spp., this attribute is combined with almost total resistance to or tolerance of available weedkillers, the problem is truly insurmountable. One plant in this group that deserves special mention is the common lesser celandine (*Ranunculus ficaria*). Superficially, this plant would seem the most intractable weed of all, possessing not only bulbils but also small brittle tubers and fairly prolific seed production. Fortunately, it is susceptible to chemical weedkillers, but, more importantly, it grows and flowers early in the spring and all but vanishes by the start of summer and thus has little impact on the growth of garden plants.

Unfortunately, treatment of any garden plant problem, be it pest, disease or weed, has come to be associated in some quarters with the use of a synthetic chemical. There have been two sides to this particular coin – the agrochemical industry for understandable reasons of vested interest has promoted the notion significantly to the exclusion of others, whereas what has come to be called the 'environmental lobby' has countered this with the advocacy of an agrochemical-free world. As with so much of life, there must be a median course and I shall approach the treatment of the problems in four ways: first, an explanation of the natural regulatory factors that operate without our intervention but which might be enhanced with it, second, a consideration of the 'prevention is better than cure' approach, third, an examination of the value of direct, but non-chemical controls, and, finally, a critique of the pros, cons and reasoned selection of chemical treatments.

Most gardeners will surely be familiar with the expression 'the balance of nature', an evocative phrase, but one that simply says that unless human beings or some major natural catastrophe intervene, no one organism in the world ever subverts another in the long term. I hope it will be apparent from what I said in Chapter Three that whilst gardeners make use of natural attributes of their plants and of the environment, the garden is far from a natural habitat. There can be no *overall* balance of nature here, and I do not propose to become involved in arguments about the feasibility or otherwise of creating so-called wild gardens. However, there are a few simple precautions that will certainly encourage natural pest control, and a few of the better understood examples of these natural processes will serve to illustrate this. Ladybirds and hover-flies or syrphids are, I am sure, fairly well known to be major influences on the size

of aphid populations. However, it may not be realised just how great this influence can be – during the three weeks of its life, a single ladybird larva may eat five hundred aphids, a syrphid larva perhaps twice as many and the latter have been observed to consume twenty large aphids in as many minutes at a single sitting. And whilst there are species of capsid bug, for instance, that are significant pests of fruit and other trees, yet other species have positive benefit to the extent of each individual capsid eating three or four thousand red spider mites.

Moreover, whilst the activity and effects of relatively large predators may be generally unappreciated, the significance of micro-organisms in controlling either pests or other micro-organisms is even less so. But it is very largely the effects of fungal, bacterial and viral diseases of pests themselves that ensure that the millions of tonnes of aphids never materialise, and it is these micro-organisms that are often the causes of the so-called population collapses that follow population explosions. Many older gardeners will recall the vast populations of large white butterflies (*Pieris brassicae*), whose caterpillars devastated garden brassicas in Britain for a few years during the early 1950s. Yet as fast as they multiplied, so they declined when the activities of ichneumon flies that lay their eggs within the living caterpillars and several bacterial and viral diseases took their toll. But, and it is a most important 'but', if we intend to adapt such natural regulatory factors to our own ends, it is essential to remember that not all the aphids are eaten by ladybirds and syrphids and not all the large white butterfly caterpillars are eradicated by parasites and predators. A residue must remain, not simply because the victim species would otherwise become extinct, but because the predators and parasites would vanish too as their food supply disappeared. This is one very important manifestation of nature's balance. Thus, if the notion of what has come to be called biological control of pests and diseases appeals to you (and I have indicated opposite the range of these tech-

Biological control methods available for garden use

Pest or disease problem	Biological control measure	Limitations
Caterpillars	*Bacillus thuringiensis* – a bacterium which is sprayed onto plants to be eaten by caterpillars and causes lethal effects in the gut	Accurate spraying is essential to ensure coverage of undersides of leaves. Not suitable for all types of plant
Whitefly (*Trialeurodes vaporariorum*)	*Encarsia formosa* – a parasitic hymenopteran that lays its eggs on the immature scale stage of the whitefly which then dies	Only feasible in greenhouses, and parasite dies out if all whiteflies are killed
Red spider mite (*Tetranychus* spp.)	*Phytoseiulus persimilis* – a South American predatory mite that eats the pest species	Only feasible in greenhouses, and predator dies out if all pest mites are killed
Dutch elm disease (caused by the fungus *Ceratocystis ulmi*)	*Triochoderma viride* – a fungus antagonistic to the pathogen. Applied as pellets of fungus into holes drilled into the tree	Difficult to introduce sufficient antagonistic fungus to eradicate the pathogen totally
Silver leaf disease (caused by the fungus *Chondrostereum purpureum*)	*Trichoderma viride* – a fungus antagonistic to the pathogen. Applied as a paste onto pruning cuts	On old trees, it is almost impossible to prevent the pathogen entering through undetected natural wounds

niques available to gardeners), you should bear in mind that if you are fully successful in your greenhouse and manage to control all of the whiteflies or red spider mites, the controlling organisms that you have employed will also die out and a fresh culture must be bought when new populations of pests find their way onto your premises from nearby.

Much damage from pests, diseases and weeds can be prevented by acting promptly, whichever direct control measure is adopted. Remove the first signs of leaf spotting on primulas, the first colonies of aphids on roses and broad beans and the first seedlings of hairy bitter-cress as they spring up among your garden plants, and you will in different ways lessen the chances of those problems becoming out of hand. For, as I indicated with reference to aphids, multiplication of most organisms is exponential – remove one and you prevent vast numbers of progeny from developing. Knowledge

of the ecology of pests, diseases and weeds will also help – I have always believed that the good gardener is a pretty good naturalist too, so a willingness to learn of such things should not be alien. Remove old prunings and other dead woody material from the garden and you remove substrates for many non-obligate biotrophic fungi such as coral spot (*Nectria cinnabarina*) – substrates that are called reservoirs of infection. Be prompt in removing dying leaves from greenhouse plants and you lessen the likelihood of grey mould (*Botrytis cinerea*) becoming established. Understand too the need of this same organism for cool, damp conditions and you will ensure that the greenhouse is adequately ventilated, especially at the beginning and end of the season when such conditions are most likely to arise. Certainly, a relatively neat and tidy garden is much more likely to be healthy than one that is cluttered with rubbish and debris.

However, you should exercise caution in your cleanliness – and use balanced judgement too. By removing hedges and replacing them with fences, you will remove the source of a number of weed problems and quite possibly the overwintering places of several garden pests and a number of disease-causing fungi. But you will also very probably remove the shelter for the beneficial predator insects such as the ladybirds and hover-flies that I mentioned earlier, the nesting places for birds and, of course, the food and homes of countless other creatures that may have no direct role, beneficial or detrimental, in your garden but which occupy an essential place in that complex of food chains and food webs that is part of 'the balance of nature'. Neither the role of natural predators nor the harm that can ensue from their destruction should be minimised.

There are several operations falling within the bounds of what may be called normal plant husbandry that have valuable roles to play in limiting the impact of pest, disease and weed problems. In general, it is probably true to say that the better a plant is grown, the more healthy it is likely to be. Just as a healthy human is more able to resist

The garden is the territory of creatures that work to one's advantage. One of the larger and more obvious of these is the hedgehog which plays a major part in the control of slugs and other pests.

infection, so the natural defence mechanisms of plants (the production of toxins to combat fungal invasion or the formation of specially thickened cells to isolate a wound, for instance) are usually more effective in a vigorous plant. It may be possible to encourage problems through 'kindness' nonetheless, and over-feeding with nitrogenous fertilisers does seem in some instances to result in soft tissues prone to fungal attack – although it must also be admitted that detailed scientific evidence for this is limited.

Attention to the pruning requirements of individual plants will ensure that dead tissues are removed promptly and also limit the air stagnation and low light levels that can be the prelude to infection. Prompt removal too of plants at the end of their productive season will minimise the likelihood of pests and disease-causing organisms obtaining a foothold, or of having a means of persisting from season to season. Careful washing and disinfection of pots, seed trays and other garden equipment should be an annual routine. And regular cultivation of the soil will ensure that pests living actively or lying dormant within it are brought to the surface where birds and other natural predators can play their full part.

The value of crop rotation has been mentioned in relation to its role in plant nutrition (p. 93), but it has an additional if lesser role to play in maintaining garden health. The theory of crop rotation for plant health is sound enough and depends on two principles – first, that many pests, fungi and bacteria are fairly specific to individual types of host plant and cannot survive on other types of plant and, second, that the period over which such organisms can survive in the soil in the absence of the correct host is very limited. There are exceptions to the first category – the grey mould fungus (*Botrytis cinerea*) and the potato-peach aphid (*Myzus persicae*) both affect a very wide range of plants – but in general it is fairly sound. There are also some organisms that can survive for very many years in the soil – the clubroot-causing fungus (*Plasmodiophora brassicae*) for over twenty years is one such – but in general, this, too, is a reasonable notion (see right). Unfortunately, much of the theory of crop rotation was established in commercial horticulture and farming where the distances between the site in which a crop is grown in one year and that to which it is moved the next may be considerable. In a garden or allotment the distance is likely to be a few metres at the most, and the conventional three-course garden crop rotation (p. 92) takes no account of the fact that many pests can fly or even that soil contaminated with *Plasmodiophora* or other organisms will be moved on spades, wheelbarrows and gardeners' boots. Crop rotation *is* a sound enough practice, but its benefits in gardens will be almost entirely nutritional.

Not all plants, even of the same species, variety or cultivar, are equally susceptible to all pests and disease-causing organisms. This arises partly because of genetic variation in the plants that confers upon them differing abilities to combat attack, but also because of variation in the pest or pathogen in its ability to counter the defences of the plant. There are, however, relatively few instances in which the pest or disease-resistance of a particular cultivar confers special benefits to gardeners, for most of the careful and painstaking plant breeding

Longevity of some common garden pests and pathogens in the absence of their host plants[a]

Pest/pathogen	Longevity (years)
Athous and *Agriotes* spp. (click beetles/wireworms)	3
Byturis tomentosus (raspberry beetle)	2
Delia brassicae (cabbage root fly)	2
Ditylenchus dipsaci (stem eelworm)	2–3
Globodera rostochiensis (yellow potato cyst eelworm)	10
Heterodera goettingiana (pea cyst eelworm)	7–8
Melolontha melolontha and other spp. (chafers)	2–6
Merodon equestris (narcissus fly)	2
Noctuid moths (various spp.) (cutworms)	2
Otiorhyncus spp. (vine weevils)	2
Phytophthora infestans (potato blight)	0.75
Plasmodiophora brassicae (clubroot)	20+
Psila rosae (carrot fly)	2
Synchytrium endobioticum (potato wart)	30+
Tipula spp. (crane flies)	3

[a] it is important to realise that these figures are based on the absence of *all* host plants, including wild species that may harbour pests and pathogens and lengthen their survival period when the appropriate garden plants are not grown

and selection needed to identify and develop these attributes has understandably been conducted on cereals and other major commercial crop species. I have, however, listed on the right those garden plants for which resistance to a particular problem may be an important reason governing cultivar choice.

The most obvious direct methods of controlling weeds are by digging them up wholesale or by severing their stems with a hoe, but these tech-

Some common plants in which resistance (or significantly high susceptibility) to a pest or disease problem may be a major factor governing choice of cultivars[a]

Plant	Pest/pathogen to which resistance is available
Antirrhinum	Antirrhinum rust (*Puccinia antirrhini*)
Apple	Apple and pear canker (*Nectria galligena*)
Apple	Sawfly (*Hoplocampa testudinea*)
Apple	Scab (*Venturia inaequalis*)
Apple	Several viruses
Carnation	Carnation rust (*Uromyces dianthi*)
Chrysanthemum	Chrysanthemum eelworm (*Aphelenchoides ritzemabosi*)
Chrysanthemum	Chrysanthemum leaf miner (*Phytomyza syngenesiae*)
Chrysanthemum	Chrysanthemum rust (*Puccinia chrysanthemi*)
Chrysanthemum	Chrysanthemum stool miner (*Psila nigricornis*)
Clematis	Wilt/dieback (*Ascochyta* sp. and other factors)
Cucumber	Leaf blotch (*Corynespora cassiicola*)
Currants	Leaf spot (*Pseudopeziza ribis*)
Grapevine	Vine phylloxera (*Daktulosphaira vitifoliae*)
Hollyhock	Hollyhock rust (*Puccinia malvacearum*)
Lettuce	Downy mildew (*Bremia lactucae*)
Lettuce	Lettuce mosaic virus
Lettuce	Root aphid (*Pemphigus bursarius*)
Mint	Mint rust (*Puccinia menthae*)
Narcissus	Several viruses
Potato	Blight (*Phytophthora infestans*)
Potato	Common scab (*Streptomyces* spp.)
Potato	Dry rot (*Fusarium* spp.)
Potato	Keeled slugs (*Milax* spp.)
Potato	Several viruses
Potato	Skin spot (*Polyscytalum pustulans*)
Potato	Wart (*Synchytrium endobioticum*)
Potato	Yellow cyst eelworm (*Globodera rostochiensis*)
Prunus spp.	Cherry blackfly (*Myzus cerasi*)
Raspberry	Cane and leaf spot (*Elsinoë veneta*)
Raspberry	Cane blight (*Leptosphaeria coniothyrium*)
Raspberry	Raspberry mosaic and other viruses
Rhododendron	Rhododendron bug (*Stephanitis rhododendri*)
Rhododendron	Rhododendron whitefly (*Dialeurodes chittendeni*)
Rose	Blackspot (*Diplocarpon rosae*)
Rose	Leaf-rolling sawfly (*Blennocampa pusilla*)
Rose	Rose powdery mildew (*Sphaerotheca pannosa*)
Rose	Rose rust (*Phragmidium tuberculatum*)
Strawberry	Strawberry powdery mildew (*Sphaerotheca macularis*)
Swede	Clubroot (*Plasmodiophora brassicae*)
Tomato	Leaf mould (*Fulvia fulva*)
Tomato	Tomato mosaic virus

[a] in many of these instances, cultivars with resistance to the problems are likely to be mentioned specifically in seed or nursery catalogues

The re-use of seedling compost is to be avoided at all costs. The tray on the left demonstrates this graphically, and contains not only re-used compost but also the fungi that cause damping-off.

niques do have their limitations. They are of very little value with many perennial weeds which may well regenerate from the smallest part of a subterranean root or rhizome, although it is sometimes claimed that repeatedly cutting off the top growth even of such deep-seated weeds as bindweed (*Convolvulus*) will weaken them in time. Most annual weeds can be controlled by hand-digging and hoeing, although it is important to bear in mind the features of weed biology that I outlined on p. 142. The weeds must be cut down or removed before they have had time to flower and seed, and some may actually need to be removed by raking if they are hoed down while still bearing flowers. And, of course, in wet weather, hoeing may amount to little more than transplanting. But if weed remains are to be disposed of, a word should be said here about the safety of composting them.

I outlined the operation of a compost bin on p. 56, and certainly all above-ground vegetative growth should rot down in almost any compost-making operation. But the ability of seeds and of rhizomes in particular to be effectively dealt with depends not on the likelihood of them being rotted but on the likelihood of them being killed by the high temperatures generated. Most seeds are destroyed by a temperature of 75 – 77 °C sustained for about ten minutes, although a few can survive 100 °C for brief periods. A properly functioning compost bin should attain a temperature of about 75 °C, at least in the centre, and should maintain this not for ten minutes but for many weeks. There should be no problem in killing weed seeds therefore if the compost is well made and preferably turned to ensure that all parts are exposed at some time to the maximum temperature. No studies seem to have been made of the tolerance of high temperatures displayed by rhizomes such as those of couch grass (*Agropyron repens*), and different gardeners have conflicting experience of composting them. Because the consequences of spreading this weed are potentially so troublesome, it may be wise not to attempt to compost the rhizomes unless you are certain of the efficiency of your compost bin – or, at least, not to compost them until you have allowed the rhizomes to dry out thoroughly first.

One of the most valuable of all garden tasks is mulching and I have dwelt on its merit for moisture retention in Chapter Four. It also has some value in minimising the transfer of fungal spores and other potentially harmful micro-organisms by rain splash from the soil onto flowers or low-hanging fruit. I am unaware, however, of any evidence that an organic mulch can actually encourage the build-up of pests, as is sometimes claimed. But it is in the control of annual weeds that mulching has one of its greatest merits. Any gardener who has ever sowed seeds will realise that each has its optimum sowing depth and that if placed appreciably below this depth the seedling may never reach the surface of the soil or compost – usually because its nutrient reserves are inadequate. It is precisely this feature that enables almost all emergence of annual weed seedlings to be suppressed by a mulch – either an organic mulch of about 5 centimetres thickness or, in the less ornamental parts of the garden, a sheet of black plastic.

When all other pest, disease and weed control measures have proved inadequate in preventing or eradicating a particular problem, the option is always open to a gardener to use a chemical treatment. There is nothing new about the chemical control of plant problems – herbicides (weedkillers), fungicides and insecticides have been in use since the nineteenth century. It must be said that, during this period, some quite astonishingly toxic and noxious chemicals have found themselves being used for one or other of these purposes, and it was the increasing and largely uncontrolled use of some of the persistent organochlorine insecticides that led the American writer Rachel Carson to publish her profoundly disturbing book *Silent Spring* in 1962. Since then, the use of agrochemicals of all types has been much more carefully regulated and those available to gardeners today are very few indeed. It is inappropriate for me to suggest whether or not any individual gardener uses a chemical pesticide or herbicide in the garden, but it is useful to appreciate the range of substances on offer, both in terms of their chemical composition and mode of action.

Fungicides and insecticides can be subdivided in various ways but almost all are relatively harmless to plant life. The only significant general exception to this is the mixture of tar oils sold primarily as a winter wash for killing overwintering pests on dormant deciduous trees, for this can damage living green tissues. Many other products may cause some harm to particular plant species or even to particular cultivars at certain stages of their growth, but guidance to this effect is given on product labels. The three most important divisions of insecticides and fungicides (and in some degree weedkillers too) in respect of mode of action are those into systemic and non-systemic or contact, into persistent and non-persistent, and into protectant and eradicant products. I should stress, however, that the three categories are not mutually exclusive.

A systemic or, as it is sometimes called, a translocated chemical is one that is absorbed by a plant and moved within its tissues. By contrast, a non-systemic substance is not or is only very slightly absorbed and remains more or less where it is applied. In practice, almost all so-called systemic chemicals are only semi-systemic – they are moved within the plant principally in the xylem but not in the phloem and thus will move upwards but not downwards – sprayed onto the leaves, no chemical will reach the roots, but drenched onto the compost around the plant, chemical should reach most parts in time. Of course, as a plant grows, so its tissue volume increases, and the concentration of chemical within is diluted. Even with a systemic substance therefore, repeat applications may well be necessary.

The likelihood of a new substance having the ability to be absorbed in this manner is a difficult one for chemists to predict – chemicals that may be quite closely related structurally one to the other may not necessarily have the same ability to be absorbed by plants. And whilst usable insecticides with systemic properties were discovered fairly early in this century, the first commercial systemic fungicide was not available until the 1960s. I have set out below the advantages and disadvantages among insecticides and fungicides that are conferred by these differing modes of action. It will be seen that, in general, a systemic product does have

Advantages and disadvantages of systemic and non-systemic pesticides

Systemic

1. Little accuracy required in spraying; can reach quite inaccessible parts of plants
2. Once spray has dried, unaffected by rain or other weather conditions
3. Usually required in very small amounts
4. Easily able to eradicate established problems, even when well-entrenched
5. By their nature, must be quite harmless to plants

Non-systemic

1. Must be sprayed precisely because only sprayed parts will be protected; significant problems likely to arise with pests or pathogens under leaves or in other inaccessible places
2. Permanently liable to be washed off by rain
3. Usually required in relatively large amounts
4. Often have only a protective action and can only eradicate problems actually on the exposed surface
5. Sometimes can cause damage to plants, especially in hot weather

major benefits as far as actual control is concerned although the fact that the chemical cannot simply be washed off the plant may be important with edible produce treated close to harvest time. Most agrochemical uses are allocated a minimal interval of time which must elapse between the application of the product and the suitability of the plant material for consumption; this period tends to be longer with systemic substances.

Systemic or translocated weedkillers fall into several different groups and include the hormone-like selective weedkillers, 2,4–D and glyphosate. Here the benefits are slightly different; glyphosate, for instance, moves in part downwards in the phloem and provides the only means of effectively eradicating deep-rooted or far-creeping perennial weeds. Because of the slowness of their uptake and movement, and hence of their killing effect, gardeners are sometimes discouraged from persevering with them, but when couch grass, ground elder (*Aegopodium podagraria*) or bindweed are major garden problems, patience is a virtue that will be well repaid.

Some truly deep-rooted perennial weeds, such as bindweed, can only be controlled effectively with the aid of a translocated chemical weedkiller which is absorbed by the plant and transported into the tissues.

A persistent chemical is one that remains in an active state in the environment (especially in the soil) for a long time before it is diluted or chemically changed to an inactive state. It was the long period over which many organochlorine compounds persist in the environment and in plant and animal tissue that occasioned Rachel Carson so much concern. Such substances (which included the well-known and once-popular insecticide dichloro-diphenyltrichloroethane (DDT)) enter ecological foods chains and are passed from predator to the prey that feeds on them. Hence, birds of prey commonly succumbed to their toxic effects a long time after the chemicals were first applied to the environment for insect control. The persistence of agrochemicals is often expressed in terms of their half-life – the period over which a given quantity of chemical degrades or is otherwise reduced to half its original amount. There is no justification for using any garden insecticide or fungicide of long persistence, and no product of any sort should be used which is not inactivated within a season – the path weedkillers such as aminotriazole, simazine, ammonium thiocyanate, atrazine and MCPA should represent the extremes of persistence and are generally recommended to keep unplanted areas free from weeds for a season. Persistence in the soil is of no real virtue in a substance that must come into contact with or be absorbed by the foliage of mature plants although it is of special merit with pre-emergence products that act by preventing the germination of weed seeds. At the other end of the spectrum are those substances that are inactivated very quickly after contact with the soil. Perhaps the outstanding example is the contact herbicide paraquat which is very strongly absorbed onto soil particles and can thus be used to kill emerging seedling weeds whilst enabling the soil to be sown or planted immediately afterwards.

The third subdivision of garden chemicals is that into eradicant and protectant and is a qualification used especially with fungicides. A protectant chemical, as its name implies, provides protection against attack, usually by being fairly persistent in the plant to which it is applied. It may function by an ability to kill an immature phase of the pest or pathogen but an inability to kill a fully mature adult or a well-established fungal growth. An eradicant chemical, by contrast, is able to kill a problem already established on or in the plant tissues – in many instances, a systemic chemical able to penetrate as far as the attacking organism has penetrated will be needed to achieve this.

It will be apparent from some of the comments that I have made already that many garden chemicals display some selectivity – they are able to exercise their effects on some types of organism but not others. Among fungicides, for instance, the well-known systemic chemical benomyl is of very wide effectiveness (it is said to be broad-spectrum) but it is appreciably ineffective against diseases caused by rust fungi or against potato blight. Sometimes such selectivity is of very positive benefit – the insecticide pirimicarb, for example, kills aphids but has very little effect on other insects and thus

Toads play a largely unseen but valuable role in pest control yet are mistakenly and unaccountably discouraged by many gardeners.

leaves such natural predators as ladybirds and syrphids unharmed. But the greatest of all applications of selectivity comes with the selective herbicides first developed for the control of broad-leaved (dicotyledonous) weeds in cereal crops. These now find immense value on lawns.

It is important, however, not only to choose an appropriate chemical for a particular task but also to choose an appropriate formulation. Most garden pesticides and weedkillers are sold in concentrated form for dilution by the user and application as spray or through a watering can. This concentrate may be either liquid or a solid (usually a wettable powder which is finely dispersed as a suspension in water). It is essential that the appropriate dilution as recommended by the manufacturers is followed. Preparing a mixture of double the suggested concentration will not double the effect and could actually be positively harmful or even dangerous because of reactions between the chemical constituents. A few garden chemicals are sold as dusts, granules, smokes or as aerosols for application without further dilution.

Both liquid and solid concentrates are usually used to make up sprays for application to foliage, although the diluted product may sometimes be used for drenching onto soil or compost or for soaking bulbs, corms or other planting material. Dusts are also sometimes applied to soil or to places where insects or other pests may be concentrated, but are also valuable for treating overwintering stocks of ornamental bulbs, corms, tubers or similar matter. Smokes are used exclusively in greenhouses for the eradication of overwintering pests or pathogens. The key both to making the correct choice of product and choice of formulation is to understand something of the way that significant information is expressed on the label. The bottle or packet of substance that you buy from the garden centre will be sold under a brand name that is simply a marketing term and usually gives no clue to the actual chemical ingredients or active product. This information is to be found elsewhere on the container in legally approved form, and I have

'Rose-clear'

TO OPEN THE SEAL

Shake well, place upright on firm surface. Pull-off measure and remove cap as instructed.

Use cutter in top of measure to cut the seal.

Locate cutter on top of bottle, push to pierce seal and twist.

Measure the required quantity.

USE THE MEASURE FOR 'ROSECLEAR' ONLY

ICI

150ml e

**makes 48 litres
(10.5 gallons)
of spray**

'Roseclear' contains 50g pirimicarb, 62.5g bupirimate and 62.5g triforine* per litre

IRRITANT   FLAMMABLE

FOR USE ONLY AS A GARDEN INSECTICIDE/ FUNGICIDE

Child resistant cap · Free measure
Unbreakable bottle

'Rose-clear'

TRIPLE ACTION SYSTEMIC
INSECTICIDE & FUNGICIDE
for
**GREENFLY · BLACKFLY
MILDEW · BLACKSPOT · RUST**

Does not affect Bees,
Ladybirds or Lacewings

ICI

'Rose-clear'

FOR

MILDEW

BLACKSPOT

GREENFLY

RUST

ICI

150ml e

**makes 48 litres
(10.5 gallons)
of spray**

'Roseclear'

TRIPLE ACTION SYSTEMIC
INSECTICIDE AND FUNGICIDE
FOR ROSES AND OTHER FLOWERS

CONTROLS Greenfly, blackfly, mildew, blackspot and rust.
PROTECTS Roses and other flowers.
SYSTEMIC Contains two fungicides to prevent and eradicate disease. Contains fast acting insecticide to kill greenfly and blackfly.
DOES NOT AFFECT Beneficial insects like **bees, ladybirds and lacewings.**
WHEN TO USE Spray every 10-14 days from April onwards or when disease first appears. Spray in calm conditions, avoiding bright sunlight.
HOW TO USE SHAKE WELL and measure out:-
  3.5ml for 2pts (1.1 litres) spray.
  7ml for 4pts (2.2 litres) spray.
  14ml for 1 gal (4.5 litres) spray.
Add measured dose to water and stir or shake. If disease is already present use double above rates for first spray. Spray thoroughly. Do not store surplus mixture.

IRRITATING to eyes

PRECAUTIONS: KEEP OFF SKIN. WASH OFF SPLASHES. DO NOT BREATHE SPRAY MIST. WASH HANDS AND EXPOSED SKIN AFTER USE. KEEP AWAY FROM CHILDREN AND PETS. HARMFUL TO FISH. KEEP IN ORIGINAL CONTAINER tightly closed, in a safe place. EMPTY CONTAINER THOROUGHLY and dispose of safely.

'Roseclear' is a trade mark of Imperial Chemical Industries PLC.
*Triforine is a product of Celamerck GmbH and Co KG.
This product is approved under the Control of Pesticides Regulations 1986 for use as directed.
MAFF No. 01826
For a FREE Guide to Pest and Disease Control send S.A.E. (at least 9"x6") to:

ICI Garden Products

Imperial Chemical Industries PLC, Garden Products, (Dept. PD), PO Box 85, Farnham Surrey GU9 7UB

Proprietary brand name

Mode of action

Type of product

Approved uses

Additional positive attributes

Frequency of use and care in use

Dilutions

Cautionary advice

Note of formal approval

Hazardous properties

Restrictions in use

Active chemical ingredients

In Britain, statutory requirements oblige manufacturers of pesticides to provide full instructions for safe usage.

The safe use of garden pesticides

- Read the label on the product when you buy it and observe the precautions printed on the label *especially* those relating to storage and the need to keep pesticides away from young children – most accidents with garden and home chemicals happen to the under-fives.
- Use only the recommended dose rates. Higher rates are wasteful, will seldom improve control and are more likely to cause plant damage.
- Two garden chemicals can be mixed together into a combined spray ONLY if there is a recommendation to do so on the product label or leaflet.
- Wash out the mixing vessel and sprayer before and after each usage. This will avoid accidental damage to cultivated sensitive plants when equipment is next used. Be especially careful when changing from herbicides to insecticides or fungicides. Rinse equipment at least three times and do not forget to wash hand lances through as well.
- When using aerosol cans of pesticides, it is essential to follow the instructions carefully. Application from too close a range or at too heavy a rate can seriously damage plants. Visible wetting of the foliage with aerosol sprays is neither necessary nor desirable.
- Do not spray in bright sunshine; this can damage plants.
- With vegetables and fruit, check if the label gives an instruction on the interval between spraying and harvesting which should be adhered to before eating treated produce. The harvest interval is recommended to allow any residues of pesticide left on the crop to decline before it is eaten.

Storage of garden chemicals

- Garden chemicals should be stored safely and securely out of the reach of children or pets, preferably in a lockable cupboard.
- As with all other domestic chemicals such as bleach and other hygiene products, they should not be stored near food.
- They should be stored in dry frost-free conditions to maintain the quality of the products, their packs and their labels.

- Weedkillers should not be kept in greenhouses because of possible risk of plant damage from spillage or from possible fumigant action.

Disposal of surplus spray solutions

- There is no point in storing surplus spray solutions because most diluted insecticides and fungicides quickly deteriorate and lose their effectiveness against pests and diseases. It is easy to forget or confuse what has been prepared.
- The best advice is to mix only enough to do the job. If there is any left over after spraying, the preferred disposal method is to apply it to uncropped areas of soil, preferably into a hole dug to prevent run-off, taking into account the possible residual effects on following crops and risk to neighbouring property.
- If this is not feasible, surplus spray solutions may be emptied safely into an outside sink drain or into the lavatory – as part of the government's approvals scheme for pesticides, garden products must be formulated in such a way that they can be safely discharged through the normal waste disposal routes in the home.

NOTE avoid disposal on areas around ponds, watercourses, marshy areas and ditches.

Disposal of unwanted garden chemicals (undiluted)

- Any garden chemical packs which have lost their labels or where the label in no longer readable should not be used.
- Solid pesticides should be left in their containers with the lid firmly closed and put into the dustbin. Glass containers will need extra protection by, for example, wrapping in plenty of newspaper.
- Liquids should be emptied safely into an outside sink drain or into the lavatory and the container should be wrapped and put into the dustbin – remember they have been formulated for such a disposal route.

indicated on a sample label over the page the way that these details are presented.

Understandably, even gardeners who are committed to the principle of using pesticides and weedkillers will need reassurance that they are unlikely to do harm to any other than the target problem. The Ministry of Agriculture (Britain) and other official regulatory bodies take seriously their responsibilities to the public and no chemical may be sold to gardeners that is likely to have special hazards attached to its normal use. Of course, all chemical substances can be harmful or dangerous if mis-used, but such factors as mammalian and avian toxicity are all taken into account before clearance for use is granted. Although in Britain labels are now legally required to provide safety information, I reproduce above the rules for safe pesticide usage as supplied by the Ministry of Agriculture.

One final point should be made. Just as species of all organisms can adapt to the pressures and constraints that are put upon them and respond in Darwinian fashion by selecting a population more fitted to survive, so pests, pathogens and to some extent weeds can adapt to agrochemicals. They may do this by the preferential selection of forms with some resistance to or tolerance of a particular chemical or even through mutating or undergoing genetic change to similar effect. Clearly, when resistance or tolerance arises in a natural population of an important pest or other problem, the consequences can be serious for gardeners and commercial growers alike. To minimise the incentive for pests and pathogens to find ways of circumventing the effects of pesticides, therefore, it is wise for all gardeners to change from time to time the chemicals they use in combating any particular problem.

CHAPTER NINE

The hardware and structural fabric of the garden

Apart from moving house, as gardeners, we can do nothing to influence the climate to which our gardens are subject, although we can certainly take measures to enhance or minimise many of the climate's manifestations. First, I shall consider both light and temperature, for the two are closely linked. To enhance or regulate visible light for photosynthesis and infra-red radiation for overall growth, we use the same structures – greenhouses, cold frames and cloches. A greenhouse can be defined as a translucent structure within which solar radiation is enhanced and plants are grown. The principle on which it works is called the greenhouse effect – the short wavelength visible light radiation passes through the translucent surface, much of it then being re-emitted as warmed longer wavelengths which cannot pass back outwards and are therefore trapped. Two materials are used commonly to provide the translucent surface, glass or a plastic, especially polythene. Both transmit photosynthetically active radiation with similar efficiency, but polythene is more efficient at transmitting infra-red. Thus, a polythene greenhouse or a polythene cloche will heat up more quickly than a glass one but, conversely, it will cool down more quickly too. Dirt on the surface can easily impair transmission by at least 40 per cent, so the glass of a greenhouse should be kept clean. Plastics often tend to attract dirt by static electricity and may be harder to clean and some are degraded by ultra-violet radiation and become brittle although certain modern formulations are ultra-violet stable.

There is much more to a garden than its plants. We must give it a physical framework and make proper use of tools and other equipment in order to garden successfully.

During the summer, however, a greenhouse will become overheated unless measures are taken to counteract this. As dirty glass impairs the transmission of solar radiation, keeping the greenhouse unwashed might seem a simple solution. Unfortunately, however, dirt excludes a great deal of visible light as well as infra-red radiation and the plants within will not only be cool but also yellow and abnormally elongated (or etiolated) as their manufacture of chlorophyll and hence photosynthesis are impaired. A much better answer is to paint the outside of the glass with a specially formulated chemical 'paint' which cuts out about 30 per cent of the infra-red and the visible light. One type of paint now available to gardeners has the merit of becoming more translucent when it is wet and so it cuts out less (about 50 per cent less) of the visible light on dull, rainy days. However, even these measures will not be sufficient to keep the greenhouse cool enough in the height of the summer and ventilation will be needed too. Vents are also necessary to admit sufficient carbon dioxide for the photosynthesis which proceeds apace during the summer and also to minimise the rise in humidity that results from the water lost through leaves by transpiration.

The best ventilation system incorporates vents capable of opening at least to 40° and positioned both in the sides (preferably close to floor level) and in the roof, the total area of vents being at least 15 per cent of the floor area. For a gardener who is away from home in the daytime, automatically opening vents are extremely useful. These contain a wax cylinder that expands when warm, so opening the vent, and returns to its original volume when cool, so closing it. Some models have a spring-loaded closer to give additional force. Set the automatic vents to open at a temperature of about 22 °C for tomatoes, cucumbers, aubergines and similar warm-climate crops or at about 17 °C for seedlings, cuttings and plants from cooler climates.

Until relatively recently, the framework of greenhouses was wood or, in very large structures, cast iron. A few manufacturers still offer wooden struc-tures, using either deal, which needs regular painting, or western red cedar, a much more durable and rot-resistant softwood, although one that still benefits from biennial treatment with a colourless preservative. Most modern greenhouses, however, are built of aluminium, and there are several important differences between this and wood as a constructional material.

A wooden frame is more flexible than an aluminium one and is therefore less prone to be damaged by gales. Wood also retains heat better than aluminium, both warming up and cooling down more slowly, and a wooden greenhouse therefore is cheaper to heat in winter. Wooden glazing bars are generally slightly thicker than aluminium ones and therefore cut down slightly the amount of light entering the greenhouse, although it must be admitted that in a small, domestic structure, this is not likely to be significant. It is generally more difficult to add to or modify an aluminium greenhouse although modern constructional advances have made this less true and there are now certainly systems that facilitate the fixing of insulating material, shelves and other additions to the inside of an aluminium framework.

Greenhouses are now available in a wide range of shapes, some seeming little more than essentially ornamental. There are, however, two features to consider when choosing the shape of a greenhouse – the amount of usable ground and bench area that the shape offers and its efficiency at transmitting solar radiation. Optimum transmission is achieved when the glass or other translucent surface is at 90° to the sun's rays. Then, about 90 per cent transmission of the incident radiation should occur (there being slight absorption by the material itself), falling off gradually until, at an angle of about 40°, there is a more dramatic decline. Obviously as the sun's position relative to the earth's surface varies latitudinally and seasonally, it is impossible for a static structure to achieve optimum transmission all of the time. A compromise is needed and easily the most efficient greenhouse shape to achieve this compromise is a dome. Approximate dome shapes using panes of glass are difficult and very expensive to construct and although they have been available for many years (and in the nineteenth century were made on a grand scale), the modern domed glasshouse is usually sold more for its visual appeal than as a horticultural necessity. And although offering a relatively large ground area, the space is extremely difficult to utilise efficiently.

The next most efficient shape is a square greenhouse with an equilaterally pyramidal roof (although I have never seen such a building), or a tunnel-like structure with a semi-circular cross-section and vertical ends. By choosing the latter option, using polythene sheet instead of glass, flattening the profile slightly and supporting the sheet over metal hoops, we produce a cheap but distinctly unlovely building used widely in commercial horticulture. Next in efficiency is a shape used extensively for garden greenhouses, an oblong structure with a ridged roof angled at about 45° from the horizontal and having glazed sides at about 60° or 70°. Unfortunately, this shape is inefficient in terms of ground area usage for there is a considerable 'lost' area of floor close to the side-walls. Among conventional greenhouses therefore, we are left with the least efficient yet easily the commonest shape, the greenhouse with a ridged roof at an angle of about 45° from the horizontal and vertical sides, either entirely of glass (described as 'glass to ground') or glazed half-way and with the lower part of wood or brick. It is very easy to erect staging in such a greenhouse and almost none of the internal ground area need be unused. The lean-to greenhouse is usually an erection similar to half of the latter pattern and built against a vertical south-facing house wall. There is merit here in that heat from the adjoining building will aid the retention of warmth in the greenhouse in winter but, even when the vertical inner wall is painted white, the reflection of strong south light into the greenhouse does not fully compensate for the lack of north light. Nonetheless, the utilisation of ground area that would otherwise be little used and the saving of valuable open garden space mean that lean-to greenhouses are attractive options for small gardens.

Recently, a novel greenhouse design has appeared that is claimed to overcome many of the disadvantages of traditional patterns. In this structure (for which east-west orientation is essential), there is a long, steeply angled south-facing roof to optimise capture of radiation from the low winter sun yet reflect away some of the excessive radiation from the high sun in summer. The shorter north-facing roof has a mirror finish within whilst the vertical north wall is painted white inside. These features are claimed to enhance the reflection of winter radiation back onto the plants whilst helping to shade them in summer. I have no personal experience of this type of greenhouse, but the principles seem sound and it will be interesting to see how widely it is adopted.

The siting of a greenhouse does not always offer many choices, but ideally any oblong greenhouse should be positioned with its long axis east–west. Although for many years the orientation used was north-south, and most older structures will be found so aligned, the east-west position offers more uniform illumination. The size of a greenhouse will usually be dictated most by space limitations in the garden as a whole and by financial considerations rather than scientific ones. Nonetheless, almost every person who owns a greenhouse wishes the building was slightly larger and a good maxim is to calculate the size that you think you need and then increase this by 50 per cent before buying. Inevitably, the calculation will be governed by the use to which you wish to put the house but there are several considerations that you might like to bear in mind.

Almost all greenhouses are used for tomatoes in the summer months and eight plants are roughly the requirement of a family of four. But the fact that the tomatoes will be planted in April and will be in place until the end of September means that you can use the space for nothing else during that time – no autumn seed sowing or cutting raising and no room for seedlings in the period between

The greenhouse is probably the most costly investment that any gardener makes. It is therefore all the more important to ensure that it is fully utilised for twelve months of the year.

sowing and hardening off in the spring. (Remember that the bench or floor area occupied by a tray of pricked out seedlings is often ten times the area occupied by the original seed tray in which they were sown and, if they are transplanted to individual pots, it can be fifty times.) The most popular greenhouse size until recently was 1.8 × 2.4 metres and, in this, the eight tomato plants could just be accommodated if both sides were used. Now, the most popular size is 3.0 × 2.4 metres and this allows for the planting of five tomatoes down one side, with 3 metres of useful bench space on which peppers or aubergines might be raised in the summer, while still leaving the bench area free at seed sowing and seedling raising time. Nonetheless, a 3.0 × 2.4 metre greenhouse is about 25 per cent more costly to heat in winter than an 1.8 metre × 2.4 metre one, so allow me to offer my compromise. For a garden of up to about 0.13 hectare (one-third of an acre), I suggest the most useful size of greenhouse is one of about 3.6 × 2.4 metres, with a internal partition dividing it into two 1.8 ×

2.4 metre units. One unit can be kept for summer tomatoes and provide sheltered frost-free space for overwintering such fairly hardy plants as fuchsias or for raising the hardiest types of winter lettuce – perhaps aided by installing movable staging for the winter only. The other unit has permanent staging (including a small heated sand bench for seeds and cuttings) and is provided with heat for overwintering more tender subjects and for facilitating early and late-season plant raising. In very cold spells in the winter or when you might wish to accommodate more than the usual number of tender plants, you still have the option of opening the internal dividing doors and admitting heat to the whole house.

Although a greenhouse is warmed by the sun, it is still by generally accepted definition an unheated greenhouse until you supply additional heat artificially. And I believe that unless you do supply artificial heat you are obtaining neither maximum use nor maximum enjoyment from your investment. There are two important considerations to make, however; what should be the source of the heat and to what temperature should the greenhouse be warmed artificially? There are four principal choices for heat generation – electrical radiant or fan heaters, paraffin burners, solid fuel burners or gas burners. The table opposite lists the main pros and cons for each, but I believe very strongly that the versatility of electricity and its relative reliability and freedom from maintenance easily outweigh all other considerations. And apart from the actual space heating function, an electricity supply allows you to operate a warmed sand bench or other propagating device (see page 163) and also to install lighting, either for your own working convenience in the evenings or the winter or actually as an additional aid to plant raising. Although tubular electrical radiant heaters fixed close to floor level have the theoretical advantage (to which I subscribe) of maintaining a more uniform temperature distribution and avoiding the peaks and troughs in temperature that a thermostatically controlled fan heater produces, on bal-

Advantages and disadvantages of different greenhouse heating systems

Heating system	Significant features
Electric fan	Jointly least expensive system to run (especially if low-price tariffs are available) Expensive system to install No routine maintenance Accurate temperature control Air movement minimises disease problems No necessity for winter ventilation because no excess water vapour Temperature subject to peaks and troughs
Electric radiant tubes	Jointly least expensive system to run (especially if low-price tariffs are available) Expensive system to install No routine maintenance Accurate temperature control No air movement so possibility of disease problems in stagnant atmosphere No necessity for winter ventilation because no excess water vapour Very uniform temperature maintenance
Natural gas	Fairly inexpensive system to run Very expensive system to install and only really practicable if greenhouse is near enough to house (a lean-to for instance) to permit use of the main central heating system No routine maintenance Accurate temperature control No air movement so possibility of disease problems in stagnant atmosphere No necessity for winter ventilation because no excess water vapour Very uniform temperature maintenance
Propane	Fairly expensive system to run Expensive system to install No routine maintenance Accurate temperature control No air movement but disease problems unlikely because of need to ventilate Winter ventilation essential because of excess water vapour produced Fairly uniform temperature maintenance
Paraffin	Fairly expensive system to run Very inexpensive system to install and independent of external power supply Daily maintenance needed Very inaccurate temperature control Slight air movement but disease problems unlikely because of need to ventilate Winter ventilation essential because of excess water vapour produced Very irregular temperature maintenance

ance the latter is probably a better choice for a domestic greenhouse. A fan heater is not only considerably cheaper and simpler to install, but it also brings about air movement within the house and lessens the disease and other problems that can arise from stagnation of the air. And, in a very hot summer, it can be used simply as a fan to aid cooling.

The winter temperature to which your greenhouse should be heated can be controlled simply by a thermostat (very simply if electricity is the source of power) but, as every 2.8 °C rise in temperature approximately doubles the amount of heat required and hence the cost, the setting should be chosen carefully in relation to greenhouse function. Excluding the specialised use of a greenhouse for growing very tender tropical plants, most gardeners are faced with selecting from the following possible temperature settings. A night-time minimum of 2 °C will ensure that overwintered tender plants like pelargoniums, non-hardy fuchsias or dahlia tubers are not damaged by frost, although assuming that the temperature raises very little above this during the daytime it will barely permit anything actually to grow. A night-time minimum of 7 °C will allow you to grow winter lettuce and enable cuttings and young plants to be maintained in good order. Moreover, by the time the sowing and pricking on of bedding plants becomes important from February onwards, the warming effect of the sun should be adequate to elevate the day-time temperature and enable them to grow quite satisfactorily. A minimum of 15–16 °C and a day-time temperature of around 17 °C (together with supplementary lighting) will be needed to grow winter tomatoes and cucumbers, a fact that makes home-grown winter salads an extremely expensive undertaking. On balance, therefore, to make the best and most economical use of your greenhouse, I recommend setting your heaters to a winter minimum of 7 °C and suggest that in general, and certainly in weather that stays cold well into the spring, this be raised to about 12–14 °C as soon as seedlings that have germinated in a warmed propa-

gator are ready for pricking on. Clearly, this timing (and the additional cost it entails) will be related to the types of seeds you sow. Pelargoniums and peppers, for instance, which should be sown in January, will work out much more costly than lobelias sown in mid-March. It is important, in addition, to choose heaters that are powerful enough to maintain the chosen temperature when that outside falls to the levels to be expected in your area. I have indicated opposite the heat output needed to maintain 7 °C over a range of average minimum outdoor temperatures. Your heater manufacturer will be able to tell you which models have the required heat output.

The cost of your heating can be reduced considerably by installing insulation in the greenhouse during the winter. From the many materials available, I favour double-skin bubble polythene film, but suggest that you choose that sold for greenhouse use, as superficially similar packaging film may not be ultra-violet stable (in other words, it will be degraded by sunlight). Double-skin bubble film cuts down heat loss by about 40 per cent, and although slightly greater savings can be made with triple-skin film, this is much more bulky and cumbersome to install. Fixing kits are available for attaching the film to an aluminium greenhouse but drawing pins are quite adequate for a wooden one. Not surprisingly, the film cuts down visible light transmission inwards as well as limiting the loss of infra-red radiation outwards, but when the greenhouse is being used for the purposes I have outlined above and plants are being maintained, rather than actually grown through the darkest winter period, no harm will ensue.

It is generally suggested that the insulating film should be removed in the spring when plants do start into growth, but the constant removal and re-erection of the film is a time-consuming task. Logic suggests that it could be left in position all year round and play a part in keeping the greenhouse cool in summer in the same way as shade paints, provided the loss in visible light is not critical. Double-skin bubble film cuts down visible light transmission by about a further 10 per cent over the 10 per cent cut out by the glass itself (see p. 156). This compares with the 20–30 per cent reduction in both visible light and infra-red transmission claimed for the best shading paints, but the diminution in the infra-red radiation reaching the interior of the greenhouse will be very much less than this if the *internally* applied insulating film alone is relied upon. In other words, interior insulating film keeps a winter greenhouse warm better than it keeps a summer one cool. Experimentally, I have actually retained insulating film in a greenhouse all year round *and* applied a shading paint outside in summer with no detectable loss in a crop of tomatoes or in the growth performance of other plants. Nonetheless, as a greenhouse cannot be disinfected properly with the film in position, removal of it in the summer is strongly recommended.

For the reasons of heating costs that I have indicated, actually growing mature plants in the greenhouse in the depth of winter is a very expensive proposition, but as plants can be simply maintained in a viable condition at such times without artificial lighting, the latter will be unnecessary for most gardeners. However, for obtaining good seedling growth in late winter and early spring from the earliest sown seeds, a supplementary lighting facility is useful and reference should be made to the adjoining table where I have indicated suitable lamp types. In addition to enabling photosynthesis to occur satisfactorily, however, light has a second important function in initiating flowering, a process mediated through the photoperiodic response of plants. I have discussed the phenomenon in Chapter Two, and gardeners wishing to use photoperiodic control of flowering will, of course, have this additional reason for needing supplementary lighting.

Before leaving greenhouses, I should mention a few minor but significant practical points. Over many years of greenhouse work, I have come to the definite conclusion that the best flooring for a garden greenhouse is gravel, laid over firmed soil. Excess water can drain away freely over the entire

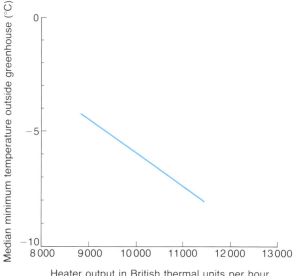

The heat output needed to maintain a greenhouse at 7 °C (in effect, frost-free) with different outside ambient temperatures. This makes it possible to judge how the cost (and cost effectiveness) varies from a cold to a mild region of the country.

Median minimum temperature outside greenhouse (°C)

Heater output in British thermal units per hour (Btu/h) to maintain 7°C inside greenhouse

Artificial lighting systems for greenhouses

Lamp type	Significant features[a]
Low-pressure sodium	Highly efficient – virtually all one wavelength (monochromatic) at around 589 nanometres Expensive to install
High-pressure sodium	Relatively inefficient – 27 per cent of energy within 400–700 nanometre waveband Expensive to install
High-pressure mercury discharge	Of variable but generally low efficiency with less than 20 per cent of energy within 400–700 nanometre waveband Expensive to install
Low-pressure mercury discharge (fluorescent tubes)	Of variable efficiency but generally around 20 per cent of energy within 400–700 nanometre waveband Relatively inexpensive to install

[a] the practical value of any lamp system is a function of various factors – principally its efficiency at converting electrical energy into light within the photosynthetically active 400–700 nanometre waveband (which will dictate its running costs), the available wattage, the form of housing and reflector and the uniformity of emission (which govern the number of lamps needed to illuminate a given area), and the lamp life

surface yet, in dry periods, it is easy to keep the floor (and the internal atmospheres) moist if you so choose. Use of a proprietary garden disinfectant will keep the base weed and pest free, and gravel looks much more attractive than concrete. Because I believe that tomatoes should be grown by ring culture (see p. 92) and because all other greenhouse plants can be raised satisfactorily in pots or even growing bags, bare beds of unsterilised soil in a greenhouse merely act as reservoirs of pests, diseases and weeds.

Greenhouse staging is available either in wood or aluminium and I shall refer to their relative merits shortly in relation to greenhouse watering systems. But staging is also available at different heights, and anyone who has worked for long hours at a greenhouse bench will appreciate the significance of this. The height is usually dictated by the manufacturers and by the height and form of the greenhouse sides, but I advise you strongly to install at least a short length of staging at a height of about 760 millimetres. This will be convenient for potting, pricking on and other greenhouse work either from a sitting or standing position (although for anyone with a specific potting bench in a potting shed at which they will work for long periods standing up, 900 millimetres is a better choice).

A cold frame is an almost essential adjunct to a greenhouse but is also a useful facility in its own right, for raising or overwintering seedlings and cuttings, for hardening off or for growing tender crops like cucumbers and melons. A well-made, insulated and correctly sited cold frame should elevate the temperature at least 7 °C above the outside ambient. Some of the considerations to be borne in mind when choosing a greenhouse apply to cold frames too – the relative merits of glass and plastic and of wood and aluminium, for instance, although the relatively better heat-retaining properties of glass and wood are magnified in such small structures. The commonest size of frame sold for garden use is about 0.6 × 1.2 metres, which should hold eight standard-sized seed trays without the plants touching the glass sides and thus becom-

ing prone to scorching, but this size is really too small for crop raising. A height of at least 30 centimetres at the front and 45 centimetres at the back is essential if tall bedding plants are not to be constricted while hardening off. The lid should be capable of being removed totally and the whole should be draught-proof yet with a facility for ventilation. It should also be capable of being firmly anchored – modern lightweight aluminum frames are very prone to damage from winter gales.

Almost all of the modern garden frames are lightweight aluminium structures and I have found not only that they are too small but also that the combination of metal frame and glazed sides renders them insufferably difficult to use effectively through a cold winter. I would urge any serious gardener therefore to look carefully at the possibility of building his or her own or of buying a larger commercial wooden Dutch light pattern frame (those with glazed covers but solid sides) before making a choice. A wooden window frame can be adapted to provide the cover for a homemade structure.

The cold frame is an essential adjunct to a greenhouse for hardening off; but if it is large enough, it can provide valuable protected growing space in its own right.

Site the frame facing south (but provide shading if you use it for rooting cuttings), and assuming that you do not use it for growing plants directly in the soil, make the base of firmed crushed cinders. Fit plastic bubble film to your frame in winter (and don't remove a cover of snow which will provide added insulation) but never assume that a cold frame is anything more that its name suggests – it will not provide complete protection from winter frost. Care should also be taken to guard a cold frame from access by very young children who could fall onto the glass; in such circumstances, a plastic-covered frame might be a wise precaution.

Cloche is a French word for bell, although the days when garden cloches were beautiful glass bells have long gone. Their purpose nonetheless is largely unchanged. Think of a cloche as a small, portable cold frame or greenhouse to give enhanced warmth and also protection from wind and rain to plants growing directly in the garden soil and you will appreciate some of its uses. Cloches can, for instance, extend the growing season for salad or other crops from early in the spring until late in the autumn, enable you to sow hardy bedding plants and vegetables earlier (an advantage of two weeks should be possible with lettuces), permit outdoor bush tomatoes, peppers or other half-hardy crops to be grown in areas where other-

The effectiveness of PVC at maintaining warmth inside a greenhouse as the temperature outside drops during a winter's night. Although plastic-covered structures warm up more quickly than glass ones, they also cool down more quickly.

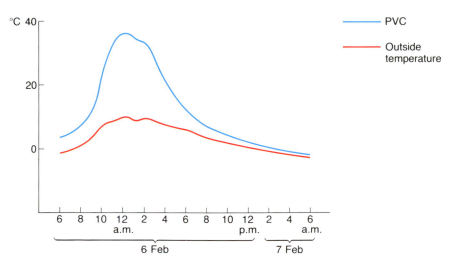

wise they cannot, protect fragile flowers from damage and, even without any plants in position, enable the soil to be pre-warmed in advance of sowing or planting. There are now many types of cloche available, some of glass, some of polythene, polypropylene, PVC or polycarbonate, and they come in a wide range of patterns, to be used either as individuals or as continuous covers (over whole rows). There are even so-called self-watering cloches that collect rainwater from the outside and permit it to trickle inwards. My choice from this vast selection is unequivocal. For enhancement of the sun's heat, durability and versatility, I believe that glass barn pattern cloches in which sheets of glass are held together by metal frames are unbeatable. They are certainly not the cheapest form of cloche, they can be fiddling to assemble initially and, of course, glass can be broken. However, for a long-term investment, a serious gardener will give thanks for his choice many times over. If for reasons of cost you do buy at least some plastic models, choose those of polycarbonate or possibly PVC.

A propagator is any small device that aids the propagation of plants, either as seeds or as cuttings by offering them enhanced temperature and humidity but on a scale that is not wasteful of heat and is therefore very cheap to operate. Perhaps the simplest propagator is a plant pot with a transparent glass or plastic cover placed on a window ledge or other position where it can make use of the sun's light and warmth. But the numbers of seeds or cuttings that can be accommodated in this way are very small and I believe the propagators based on standard sized seed trays with plastic covers offer much the most versatile system. These are small enough to stand on a window ledge and yet allow several different types of seed or cutting to be sown or struck simultaneously. But it is essential that the cover has some means of allowing additional ventilation if young seedlings and rooted cuttings are not to succumb to damping-off diseases.

Some seed-tray sized propagators have an electrical heating device in the base and this will certainly improve germination and assist young seedling growth (there is now some doubt regarding the specific merit of raising root and soil or compost temperature in the absence of raising that of the air also but as the air above the compost is warmed in these systems, the benefits are evident enough). Nonetheless, although such a propagator might be beneficial in a house where the ambient temperatures are particularly low, it has the major disadvantage of being unusable for further seed germination until the young seedlings from the previous sowing have been pricked on. A much better system is to use a propagator in which the source of heat is separate from the actual seedling container. Low-voltage equipment is available consisting of a flat heating mat on which seed trays are placed. But the most effective system of all is appropriate for a greenhouse large enough to be able to devote about a metre length of staging specifically to the purpose. This is a small sand bench which comprises a wooden frame, about 15 centimetres deep, with a plywood or hardboard base. Within this is placed about 12 centimetres of sand on which a heating cable is laid. A thermostat is fitted to the cable and the bench then filled with sand. Standard seed-tray propagators with covers may be placed on the surface of the sand and removed or replaced without disturbing the heating elements. A bench of dimensions approximately 1 metre × 0.5 metres will accommodate four standard seed trays. The running costs are very low, but it is essential to use cable and connectors designed for the purpose, to note the area of bench for which the cable is intended and to lay it in the pattern recommended by the manufacturers, ensuring particularly that there are no cross-overs.

Before leaving the subject of propagators, I should say something about the relative merits of plastic and clay pots, for although the plastic pot has now become almost ubiquitous because of its cheapness, lightness and self-evident greater robustness, it should not be considered a straight substitute. The clay pot does still have its uses. Its heaviness can be a positive advantage, for some plants (many house plants, for instance) are top-heavy, and if they are placed in plastic pots they may actually need to be over-potted (placed in a volume of compost and size of pot too great for their needs or their good) if they are to remain stable. Even more importantly, clay is permeable to moisture and air where as plastic is not. Thus, although the compost in a clay pot dries out more quickly, it is much less likely to become water-logged and the roots much less likely to rot.

Not so many years ago, garden watering meant a length of filthy black rubber hose-pipe and a galvanised iron watering can that even when empty was no weight for the faint-hearted. Times have changed to the extent that we are now faced with a bewildering array of equipment that encompasses microprocessor-controlled fully automated pop-up irrigation for the entire garden, hidden plastic lawn edgings that spring to life and spray water over borders at the touch of a button, and sophisticated trickle watering devices for the greenhouse. In effect, your garden watering can be as complicated and automated as your inclinations and bank balance allow. Both my inclinations and my bank balance take very much a median course and I shall concentrate therefore on equipment that I have found either indispensable or extremely good value in respect of its time and labour-saving benefits.

Before buying any watering equipment, it is sensible to check two matters with your local water authority. First ascertain their regulations for the use of hose-pipes and unattended watering equipment. In some areas, a permit or licence is required (sometimes with payment), whereas in others watering equipment may be used in a domestic garden for non-commercial purposes. In a few parts of the country, unattended equipment is never allowed and, of course, during periods of national water shortage, temporary restrictions may be introduced. In passing, I should mention that in the many old gardens still having a well, manual abstraction of well water is normally permissable at all times although use of electric pumping is likely to be controlled. And check, of course, if you may remove water from a stream or river that flows through your garden – you may own the stream bed, but you do not always own the water. Second, ascertain the mains water pressure in your area for this can affect the functioning of certain types of sprinkler and, if the pressure is high, many types of tap and hose connector will leak. Within Britain, mains pressure varies from about 30lb/in^2 (211 kg/cm^2) to about 130 lb/in^2 (914 kg/cm^2).

The core of any watering system is the hose and its connectors and I urge you to decide at the start which of the leading manufacturers you wish to patronise, for the equipment of one may not fit that of another. As there are few things more annoying than a leaking hose and leaking joints, my preference is with modern double-wall PVC reinforced hose which is as flexible as the non-reinforced type but much more durable. If cost is a major consideration, buy double-wall non-reinforced, but avoid cheap single-wall hose which can burst at pressures as low as 100 lb/in^2 (703 kg/cm^2). I also prefer the so-called automatic or snap-lock fittings and, most importantly, a screw thread connection to the tap. I realise that this may be impossible if the connection is to be a kitchen tap, but the problems consequent on an indoor connection coming adrift are such that you really should make a major priority of having an outside threaded tap installed. Again, to minimise the likelihood of leaks, buy a hose of the maximum length you are likely to need, rather than connect several short lengths. Store it on a through-flow reel – once used, these really do join the list of indispensables. And, of course, all of these comments are especially valid in areas where water pressure is high. If you plan to have several items of watering equipment and have a large garden, it would be worthwhile buying a hose-end connector that automatically shuts off the water supply when it is removed from the appliance – this avoids a long walk back to the tap or, the alternative, the probability of a very cold shower. Flat hose has become widely available in recent years and offers the appeal of needing less storage space, but is considerably more expensive than conventional hose and has the major disadvantage of having to be unwound fully before use.

For most gardeners, the principal item of watering equipment will be a sprinkler – a device to be left unattended for spraying water over a large area. There are four main types available. The static sprinkler, usually anchored into the ground with a spike, offers a circular spray pattern up to 6 or 8 metres in diameter, usually with no facility for adjusting droplet size. Rotating sprinklers usually have two or three whirling arms, from each of which water sprays to give circular coverage up to about 20 metres diameter. Most are crudely adjustable for coarseness of spray. Oscillating sprinklers have a single perforated bar that swings back and forth to give a more or less oblong pattern of coverage to about 16 × 13 metres, usually with the facility for limiting the coverage to one side only if required. Pulsating sprinklers incorporate a principle found widely in commercial irrigation systems. There is a single narrow-spray of water which thus makes use of the full pressure to gain distance, while the whole head moves slowly in a full or part circle to cover a large total area of up to about 26 metres diameter. Bear in mind that the distances of spray that I have indicated will fall significantly if the mains pressure drops, as it may well do when everyone is watering their gardens at the same time in the height of summer. Bear in mind that few gardens or garden features are circular and any system giving a circular spray pattern will inevitably have to water a fairly large proportion of the total area twice in order to ensure that all is covered. My strong preference is for the versatility of adjustment and fairly uniform rectangular spray pattern of oscillating sprinklers even in a small garden. However, in a very large garden (say one with at least 0.1 hectare (0.25 acre) of lawn) or where there are large areas of tall herbaceous plants to water, a pulsating sprinkler mounted on a tripod would be a wiser choice. Adjustment to give fine or coarse spraying is of little advantage with garden watering systems although it is very important with hand-held chemical sprayers (p. 149).

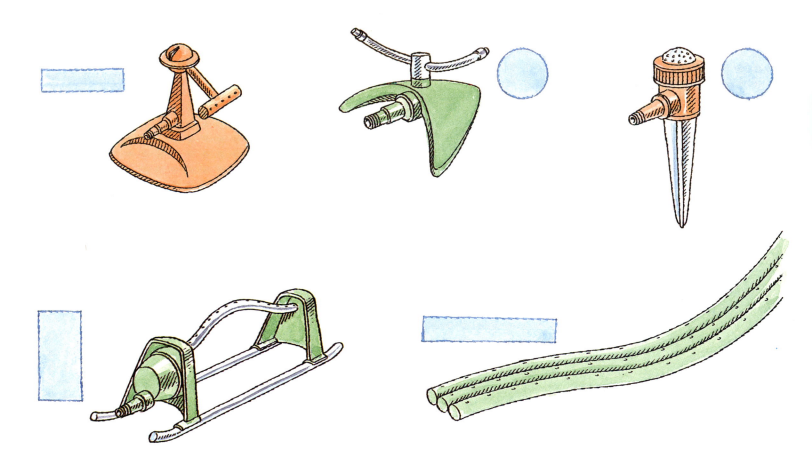

Garden sprinklers are now available in a wide array of styles capable of reaching a defined area of ground with a particular pattern of spray. The more sophisticated sprinklers are adjustable both for size and shape of the area to be covered.

An alternative to a movable sprinkler is a permanent or semi-permanent garden watering system. The technology of permanent watering systems was developed in the United States where every garden and certainly every golf course worthy of the name has a network of buried pipes attached to concealed nozzles that emerge from holes in the ground and can be finely adjusted to give spray coverage of arcs and circles of varying sizes to irrigate even very irregular areas uniformly. One limiting factor is that the devices are usually installed on a lawn (although they may spray well beyond it) to prevent loose soil from falling into the nozzles and, unless installed inconveniently *above*

soil level, they cannot be fitted into beds and borders. But the second disadvantage is that the very considerable outlay is unlikely to be worthwhile in the climate of Britain unless you live in a very dry area of the country and are often away from home for considerable periods in summer. Semi-permanent trickle irrigation systems can also be installed outdoors but can be time-consuming to lay (see below). Nonetheless, within borders and shrubberies, perforated hose-pipes do offer an attractive alternative to a sprinkler. Flat perforated hose of several types is available and may be laid (or, in one system, even buried) in the border or vegetable garden early in the season and connected to the mains when required. Depending on the type of hose used, either a gentle spray emerges upwards through numerous small holes or water seeps out through tiny pores to ensure that the soil is moistened even beneath large-leaved plants such as dahlias or brassicas.

The elevated temperatures of the greenhouse present special difficulties in maintaining plants with an adequate water supply. There are, however, several possibilities instead of interminable hand-carried cans, although all will be rendered less laborious the nearer to the greenhouse that a tap can be installed. But as it is usually tomatoes that present the greatest demand for greenhouse water, I cannot urge too strongly the merit of growing them by the ring culture system (p. 92). I have left tomatoes with no attention for ten days in the height of summer on well-watered ring culture beds and experienced only the relatively minor inconvenience of some fruit having split. For other plants, the main possibilities are trickle irrigation, overhead irrigation or capillary watering devices.

A trickle (or drip) irrigation system comprises lengths of narrow plastic tubing, each ending in a nozzle through which water drips, the supply coming originally either from the mains through some flow-restricting device or from a reservoir. Each drip nozzle supplies one plant, one pot, one growing bag or other unit. Years of experience in commercial as well as garden greenhouses convinces me that a poor trickle irrigation system is a most fiddling and frustrating tool for the amateur gardener. You must buy the best and you really do require a mains water supply as the reservoir systems very soon need refilling in the height of summer. But then if you do have the time and patience to erect a mains system carefully, it can be invaluable for watering not only plants in the greenhouse but also growing bags or a collection of plants in containers on a patio. And it is most useful of all when the owner is away (and no friendly neighbours are available). Overhead or mist irrigation systems spray water downwards from nozzles attached to the greenhouse roof, often

The water requirements of plants vary enormously. Even within a small area of garden, the moisture content of the soil will differ widely, the soil close to the wall being particularly dry.

All greenhouse irrigation systems, whether misting (left) or drip (right), depend for their efficiency, quite simply, on preventing the very small nozzles from becoming blocked.

using similar pipework and supply systems to trickle devices. In general, I find a trickle system of greater value as it is more precise – a mist system can fail adequately to water the soil or compost beneath large-leaved plants. And in a greenhouse with a mains electricity supply, a mist system must, of course, be used very carefully. Nonetheless where a gardener is sufficiently interested in striking large numbers of plants from cuttings such systems really come into their own and can be installed over a sand propagating bench.

Capillary watering systems are much less complex than trickle or overhead apparatus and they depend on the ability of compost in a pot to take up water through the draining holes in the base from some external source. Believing that simplest is generally best, my suggested capillary watering system comprises a shallow aluminium tray to fit on top of your greenhouse staging and in which a sheet of capillary matting is placed, one end of the matting overlapping one end of the tray and dipping into a reservoir trough – complete kits can be purchased ready for use. My experience has been that polyester matting 2–3 millimetres thick gives the best results, water rising from the matting into 9 centimetre diameter pots of peat-based compost by up to 5 centimetres. But capillary matting has its drawbacks. Algal growth will thrive on the matting and can be extensive and unsightly. It is best combated by twice-yearly watering of the matting with an algicide. Because there is no drainage from the pots, chemicals from the fertiliser (which should still be applied by watering can from above) can build up to damaging levels. Nitrates and potassium salts are especially important and can increase the salinity of the compost so much that the plants may actually have difficulty in taking up adequate water osmotically. It is sensible to water thoroughly from above therefore three or four times during the season to clear this problem. During the winter, the danger of overwatering and so bringing about disease problems is high and it is wise to revert then to limited hand watering. Using pots of different sizes, containing different types of plant and/or compost can result in uneven watering, and solid lumps of peat or other impurities in the compost can easily cause the drainage holes to become blocked and so impede the capillary uptake. The flatter base of clay pots gives very much better contact between compost and mat than is possible with a plastic pot and compost in a plastic pot suffers a much greater likelihood of becoming waterlogged and stale (see p. 164). And, finally, it must be said that some plants consistently fail to flourish when their compost is kept permanently moist, be they in clay or plastic containers; I have never succeeded well with pelargoniums on capillary matting for instance.

Any garden watering system that is connected via a hose to the mains water supply is likely to be amenable to some simplified automation. There are three widely available options, all of which screw onto standard tap threads and into which hose-end connectors can be snapped: a battery operated and micro-processor-controlled watering computer, a clockwork watering timer and a clockwork water measure. The watering computer allows the flow to be turned on and off for pre-determined and different periods of time once a day over a seven-day repeating cycle. It can be connected to a soil moisture sensor that overrides the pre-set mechanism if the soil is already moist. This device only really justifies its fairly high price if used in connection with some permanent or semi-permanent watering system – it is ideal with a mist propagating bench for example. The other two types of device turn off the water supply but cannot turn it on and are very useful therefore when used with a garden sprinkler turned on late in the evening or before going out. Because the water flow can vary so markedly with fluctuations in mains pressure, a control that turns off after a pre-set volume of water has been delivered, rather than after a pre-set time, makes more sense.

Devices claiming to measure soil moisture content and thus to make plant watering more meaningful have become widely available recently, for use both outdoors and with house plants. However, accurate determination of the moisture content of soil is very hard to achieve for the methods depend on actually removing the water from a soil sample. This is not only time-consuming but also difficult to do meaningfully – the methods used (the temperature to which the soil is heated or the pressure to which it is subject, for instance) dictating how much moisture is driven off. A point must therefore be selected beyond which it is decided that the remaining moisture is of no consequence for plant life. It is quite impractical for any of these accurate methods to be adopted by gardeners and the devices actually available are extremely crude. They either indicate the relative wetness of the soil by a change in the colour of a chemical indicator or by measuring a change in the electrical resistance of the soil. It is highly arguable whether these add anything to the information that can be derived from feeling the soil or compost with your finger and thumb. The coloured indicators for use with house plants do serve as useful and inexpensive reminders although it must be borne in mind that the actual water requirements of house plants vary widely. I cannot be convinced that the crude moisture meters are of any value except perhaps where they are integrated in some automatic watering system where they can be used to override a pre-set watering programme if the soil or compost is already very wet.

Whilst not related directly to the measurement of soil moisture, it is pertinent to mention here the measurement of soil pH and also of soil nutrients, for small so-called pH meters are often advertised and sold with moisture meters. Like moisture, soil pH is exceedingly difficult to measure, and even harder to measure accurately. The small meters sold for garden use do not actually measure pH at all but measure instead a small electric current set up in the soil between two dissimilar metals. This can sometimes give an approximation of pH, but on some soils the disparity is considerable (see the table above). Far cheaper and more accurate are pH testing papers impregnated with a chemical indicator that changes colour in response to pH differ-

Comparison of three methods of measuring soil pH[a]

Soil type	Professional soil laboratory	Garden 'pH meter'	Triple indicator paper
Sandy	4.76–4.83	3.0–5.0	4.50–5.00
Coarse loamy	4.95–5.22	4.5–5.0	4.75–5.00
Fine silty	6.55–6.75	6.0–6.5	6.50–6.75
Fine loamy	8.00–8.30	7.0–8.5	8.00–8.50
Clayey river alluvium	6.14–6.35	5.0–7.5	6.00–6.25

[a] five readings were made in water on samples collected from between 5 and 15 centimetres below the soil surface and bulked or collected together. The garden meter did not permit accuracy of readings greater than ±0.5, whilst the indicator papers, although graded on 0.5 pH units, enabled interpolation to be made to ±0.25 units

ences – those using a combination of three different indicators are the most accurate.

Most soil nutrient testing kits are even more inaccurate. The identification and quantification of chemical nutrients in the soil is exceedingly complicated, particularly as the presence of other chemicals with which they may be combined confuses the analytical procedures. The small garden test kits that rely on the colour change of chemical dyes can at best give a crude indication of the relative amounts of certain major nutrients, especially phosphorus and potassium. They are most misleading in their measurement of soil nitrogen which fluctuates widely under the influence of temperature and rainfall and other factors, almost on an hourly basis. In general, sound husbandry and regular fertiliser applications, coupled with the deficiency symptoms that plants themselves display when short of nutrients, should be adequate to ensure that gardeners do not need to attempt soil nutrient measurement. Nonetheless, recently, a technique has been developed for commercial horticulture that enables the nitrogen requirements of plants to be gauged from a determination of the nitrogen present in their sap.

All gardeners will have need for at least one, and probably two or three sprayers. Even if not used to apply pesticides and weedkillers, a spray with water will be helpful in maintaining the damp atmosphere required by many house and greenhouse plants. Numerous types of sprayer are now available, ranging from simple hand-held compression appliances to motorised knapsack sprayers for orchards. Sprayer size and volume will be dictated by your needs, but the delivery system of all is the same; a small hole in a nozzle provides the orifice through which the liquid emerges and it is here that much of sprayer technology is centred. The cheapest sprayers have poor-quality nozzles that deliver a spray in the form of a hollow cone – there is a dry area in the centre of the spray pattern. They also have nozzles that cannot be adjusted for droplet size. The former fault is an inconvenience, meaning that great care must be taken to ensure that all parts of the plant are treated, but the latter is technically more significant. Droplet technology is a highly complex subject, relating droplet size, specific gravity of the liquid, electrostatic and other forces, volatility and other factors to the efficiency with which a surface at a given distance from the orifice is covered. Most of these factors need not concern us here, but it is useful to appreciate the differences in behaviour of large and small droplets. The droplets of liquid from an aerosol, the smallest of those with which gardeners are likely to be familiar, have a diameter of less than 50 micrometres (μm). Above this come mist (51–100 μm), fine spray (101–200 μm), medium spray (201–400 μm) and coarse spray (above 400 μm). As may be

calculated from the tables below, whilst the total volume of liquid delivered is the same in each size class quoted, the smaller the size of droplet, the more uniform is the coverage of the surface and the greater the likelihood that any particular place (occupied, for instance, by a fungal spore or an aphid) will be struck. But against this merit of very small droplets must be weighed the fact that their surface area is very much greater (and losses from evaporation therefore very much higher) and also that they are much more readily blown by the wind – obviously, the longer a particle lingers in the air, the more likely it is to be blown away. On balance, the best maxim for gardeners to follow is to chose a sprayer with an adjustable nozzle and use the finest spray that is possible without drift in the wind becoming noticeable.

The types of hand tool now available to gardeners are extremely numerous and to describe the principles behind them all would take most of the book. Many, like the spade and the hoe, have been used in one form or another for centuries, others are recent inventions that have not yet stood the test of time. No gardener of any experience needs telling which hand tools he or she requires and it is also self-evident that, in general the more you pay, the better and more durable will be the tool. But there are some overall technical considerations worth bearing in mind. Tillage tools and cultivators comprise a working end of either stainless or carbon steel and a handle, or shaft, of lightweight

Theoretical droplet density if 10 millilitres of liquid is sprayed uniformly over a flat surface of 1 cm²

Droplet diameter (micrometres)	Number of droplets per square centimetre
10	1909900
20	238700
50	15300
100	1900
200	240
400	29.8
1000	1.9

Time taken for droplets of different sizes (with a specific gravity of 1)[a] to fall to earth in still air

Droplet diameter (micrometres)		Time to fall from a height of 3 metres (seconds)
1		101160
10	aerosol	1014
20		252
50	mist	40.5
100	fine spray	10.9
200	medium-coarse spray	4.2
500		1.65

[a] the specific gravity of water

A neat lawn edge is one of the most important cosmetic features in a garden. A purpose made, lawn-edging knife is essential to achieve this, but one should remember to slope the edge slightly so that it does not crumble.

The push hoe is perhaps the most valuable weed-controlling tool of all; and this modern alternative, the swoe, enables you to weed around the back of established plants without causing them harm.

metal such as aluminium, of plastic or of wood (usually ash), sometimes sheathed in plastic. Wood is much to be preferred for the handle of a tool like a spade or a fork that must have the strength to act in part as a lever and also take a measure of flexing. For a cultivator, like a hoe or a rake, the weight of the long handle is the major criterion and most of the lightest modern cultivators have tubular aluminium handles. There are many other features, like length and angle of the handle and form of the hand grip, that can only be judged by each individual. Never buy a hand tool therefore without handling it thoroughly first in the shop or garden centre.

There are two other aspects to choosing a cultivator. Whilst many types (the conventional Dutch hoe, for instance) depend on the gardener employing a pushing action, others rely on the ancient draw principle – the tool is pulled through the soil towards the user who remains in a more or less upright position, a method that is claimed to lessen the effort required, minimise the likelihood of back injury, give more uniform cultivation depth and lessen the necessity for walking over already cultivated ground. The draw system is much more widely used and popular in some countries than others, and British gardeners have been more

reluctant than many to adopt it. Before buying a range of tools employing this principle, therefore, it would be sensible to try it out with one. But a second consideration with cultivators relates to a more recent development – the detachable head system. Here, the same handle serves for hoes, rakes and other cultivators, the heads of which are easily interchanged. The merit is the ability it offers for you to own a wide range of tools at lower cost and, possibly, to lessen storage space. The disadvantage is the inabilty to swap very quickly from one tool to another; or, of course, for two people to work in tandem. Perhaps owning two handles would be a simple compromise.

Modern spades and forks are made of steel and you will see descriptions applied to them such as stainless, carbon, high carbon, forged, hot forged, solid forged, one-piece forged, cast, tempered, heat-treated, rivet-strapped, open socket, solid socket, solid necked, tubular shafted, hammer finished, epoxy coated, or polished. I doubt whether one gardener in a hundred knows the significance of all these apparently imposing attributes, so I have attempted to clarify in the table overleaf the meaning of some of the terms met with most commonly.

Among cutting tools, the most important con-

171

Meanings of some of the terms commonly applied to garden forks, spades and other hand tools[a]

Carbon	Steel, with higher carbon content than mild steel; suitable for heat treatment (tempering)
Cast	Molten metal poured into a mould and then allowed to set
Epoxy coated	A powder paint coating
Forged	Billet forged – shaped under a press
Hammer finished	A paint finish with a mottled effect that hides surface imperfections
Heat-treated	Steel, structurally changed and hardened by heating
High carbon	Steel, suitable for heat treatment (tempering)
Hot forged	Heated then forged (the most common forging method)
One-piece forged	As forged
Open socket	A socket made from pressed steel – the shaft is visible
Polished	Stainless steel finished to a mirror-like surface; usually achieved by craftsmen
Rivet-strapped	Two forged strips of steel at the front and back of the shaft to which they are riveted
Solid forged	As forged
Solid necked	A solid steel portion of the shaft just above the foot area of the tool
Solid socket	A forged socket almost entirely enclosing the shaft
Stainless	Steel, carbon steel with high chromium content
Tempered	Hardened by heating followed by dousing
Tubular shafted	A steel tube rather than ash wood shaft

[a] for those sufficiently interested in the technology of garden hand tools, British Standard BS 3388:1973 'Specification for forks, shovels and spades' (among other British Standards) will provide copious information

Always use sharp secateurs and always cut to just above a bud and at a sloping angle to avoid leaving a dead stub.

To achieve the very best results from your lawn, a cylinder mower is unrivalled, as no other mowing technique provides such a smooth, velvety surface.

sideration relates to pruners or secateurs of which there are two main types, employing very different principles. One type, a nineteenth-century invention, has two blades working together on a scissor system. This slices very effectively through fairly soft stems without harming them and is ideal for rose pruning, which accounts for the bulk of pruning activity in most gardens. A more recent development is the anvil pruner which has one blade cutting onto a flat edge. The cutting strength of this is greater and it is much better for cutting hard, woody tissue in which scissor pruners can easily be twisted. Soft stems, nonetheless, can readily be crushed with the non-cutting face of the tool. On balance, a pair of high-quality scissor action pruners combined with a pair of long-handled anvil action lopers should cope with most situations.

Ever since Edwin Budding, a textile engineer, employed his knowledge of the machinery used to trim the nap of cloth and invented the cylinder lawnmower around the year 1830, mowing machines have been essential items of gardening equipment. Today, they account for a large proportion of the annual expenditure on gardening, and whilst the cylinder machine, based very closely on Budding's principle but usually motorised, is very much still with us, it has been joined by powered wheeled or hovering rotary mowers.

Cylinder mowers have sharp blades that cut with a scissor action as the cylinder rotates giving a clean, straight cut. The uniformity (or, as it is generally called, the fineness) of the mown lawn surface is a feature of the number of grass leaves that have been cut and the uniformity of length to which they have been cut. These features are functions of blade sharpness and of the number of cuts per unit area (governed by the number of blades and the speed at which they turn relative to the forward motion of the entire machine). Lawnmowers for domestic use generally have five or six blades, although those used for bowling greens and golf tees may have up to ten. A five-bladed machine, hand propelled, will give about 40 cuts per metre, a six bladed petrol-driven model, about

80. Electric cylinder mowers generally have fewer blades than petrol-engined machines but the blades rotate more quickly.

Rotary mowers usually have a single, centrally pivoted blade, sometimes of metal, but in smaller machines of nylon or other plastic. Some types, generally called orbital mowers, now have a spinning plastic disc instead of a metal blade, with or without a flail-like cord attached. But all rotary mowers rely on their speed of rotation (between about 2500 and 4000 rpm) to slash the grass; they do not cut it in the accepted sense of the word. Assuming an average walking speed with a mower of 5 kph (about 3 mph), the average number of cuts (or slashes) per metre with rotary mowers ranges from about 60 to about 80; comparable therefore with cylinder models. But it should always be remembered that the mowing action is fundamentally different and a rotary mower, no matter how fast it rotates, will never give your lawn the velvety surface so beloved of greenkeepers.

Besides the fineness of cut, another important aspect of lawn care is the speed with which a given area of grass can be mowed. This depends almost entirely on the width of the cut. The tables above give cutting times and recommended mower cut widths for different sizes of lawn.

Although the lawnmower is the commonest powered garden appliance, it is no longer the only one. Hedge trimmers, lawn rakers, grass trimmers (often called strimmers, although this is in fact a brand name), compost shredders, cultivators, chain saws, lawn edge cutters, weeders and miniature cultivators are among the commonest. I have been unimpressed with the latter three but find all of the others useful – although obviously their general value is related to the size of garden, length of hedge and so forth. The best hedge trimmers are those with two reciprocating blades but, other than on very extensive hedges, single-bladed models of lengths about 35 centimetres (14 inches) will be quite adequate. Electrical grass trimmers with an automatic or semi-automatic nylon line feed are the easiest type to use, but if you plan to indulge in

Lawnmower widths and cutting times

Width of cut (centimetres)	Area mown in 1 hour[a] (square metres)
30	1230
35	1450
40	1670
45	1938
50	2205

[a] assuming a walking speed of 5 kph (3 mph) and 5 centimetres overlap between cuts

Area of lawn and recommended lawnmower widths

Area of lawn (square metres)	Recommended mower width (centimetres)
Under 100	30
100–200	35
200–400	40–45
Over 400	50

brush rather than grass cutting, a petrol-driven machine with interchangeable cutting heads is essential. Powered compost shredders are a boon, for they not only minimise the amount of garden debris to be bagged up and carted away, but are valuable in allowing a proportion of woody matter to be incorporated in the compost, thus permitting it a better textural blend (see p. 56). An electric machine of about 1000 watts is just about adequate to cope with the material from a garden of up to about 0.1 hectare (0.25 acre). Only if you have a large vegetable garden is a powered cultivator worthwhile and, even then, they have drawbacks (p. 51). Choose a petrol-engined machine of at least 25 horsepower with variable handle positions and with a good range of attachments; but do be sure that you can handle it in operation. Cultivators are big, heavy, rugged items of equipment and are often difficult to steer.

All gardens have boundaries and, in all except so-called open-plan housing estates, the boundaries are marked by some vertical structure – a fence, a wall, or a hedge. These structures give privacy and can be used as important contributors to the overall aesthetics of the garden. But they also provide most gardens with much of their shelter, and it is their relative merits in this respect that concern me here.

When we speak of shelter in a garden, it is shelter against the wind (see Chapter Three) with which we are concerned, but a related consideration is shade against the sun, which may be a much less desirable result. Lessening the force of the wind is not just a matter of erecting a barrier in its path. This serves merely to deflect the wind upwards so resulting in an area of low pressure on the leeward side and bringing about downward air currents which may actually cause more damage than the original wind. Lengthy experiments on the design of windbreak barriers for the protection of commercial fruit tree orchards and forest plantations have revealed that the most efficient windbreak is one that is 50 per cent permeable. It has been found that a barrier of this type will reduce wind speed and force for a distance to the leeward of thirty times its height, although the maximum benefit occurs only for ten times its height. Therefore, a barrier of 2 metres high will give significant protection to a garden 20 metres wide and will be adequate for most domestic situations. It is in choosing the type of 50 per cent permeable barrier that the greatest difficulties lie. I have therefore summarised on the right the relative merits of the various wall, fence and hedge options.

Like any other recreational pursuit, gardening does not happen by itself or by accident. We require the appropriate tools for the various tasks, and I hope that this account will enable you to choose them meaningfully and thus use them more wisely. But if this is true of tools, surely it is true of the entire subject. For those of you who began this book with apprehension and foreboding, I hope you are now reassured to discover that science is not after all the daunting subject that you may have shunned at school; it is simply the key to understanding, which in turn is the way to success. And I believe that through this success your horticultural activities will bring you even more enjoyment and fulfilment in the future.

Relative (non-aesthetic) merits of different types of garden windbreak

Walls	Quickly erected
	Occupy little space
	Require little maintenance
	Relatively the most expensive
	Difficult to achieve 50 per cent permeability without having gaps or holes so large that they adversely affect privacy
Fences	Quickly erected
	Occupy little space
	Limited life; posts are likely to rot within fifteen years
	The commonest types of overlapping soft-wood panels are barely permeable and even if adequately secured (use posts sunk at least 60 centimetres and preferably 1 metre in the ground with diagonal bracing against every second post for standard 2-metre high panels), the lee-side turbulence can result in large quantities of autumn leaves collecting in your garden
	Slatted fences are adequately wind permeable but this benefit can be lost if evergreen climbing plants are grown over them; deciduous climbers are less of a problem for the permeability in winter will be little affected and damage from gales in summer is unlikely
Hedges	Very slow to establish
	Occupy considerable space
	Laborious to maintain; most will need at least two clippings per year
	Harbour beneficial insects but also harmful pests and some disease-causing organisms
	Can cause the soil nearby to become very dry and impoverished
	Generally adequately permeable, any lack of permeability in evergreen hedges in winter being compensated for by their considerable flexibility

Page numbers in *italics* indicate illustrations

ACKNOWLEDGEMENTS

Photographs in the book are in the copyright of Stefan Buczacki with the exception of those on the pages shown below which are reproduced by kind permission.

p.11, Ronald Gray; p.24 *top*, University Botanic Garden, Cambridge (P.F. Yeo); p.31, U.B.G.C. (Warner); p.32, U.B.G.C. (P.F. Yeo); p.49, Alan Bloom Garden; Marianne Majerus Garden Picture Library; p.53, Photos Horticultural; pp. 57, 64, U.B.G.C. (P.F. Yeo); p.74, Photos Horticultural; p.85, U.B.G.C.; p.89, Hugh Palmer; p.91, Martin Walters; p.112, A-Z Collection/Photos Horticultural; p.113, Hugh Palmer; p.114, Statens Konstmuseer; pp.133, 158, 167, U.B.G.C. (W.H. Palmer).

The illustration on p.152 is reproduced courtesy of ICI Ltd.